MW00452283

Sensory and Instrumental Evaluation of Alcoholic Beverages

Sensory and Instrumental Evaluation of Alcoholic Beverages

Hildegarde Heymann
Susan E. Ebeler

AMSTERDAM • BOSTON • HEIDELBERG • LONDON • NEW YORK • OXFORD • PARIS
SAN DIEGO • SAN FRANCISCO • SINGAPORE • SYDNEY • TOKYO
Academic Press is an imprint of Elsevier

Academic Press is an imprint of Elsevier
125 London Wall, London EC2Y 5AS, United Kingdom
525 B Street, Suite 1800, San Diego, CA 92101-4495, United States
50 Hampshire Street, 5th Floor, Cambridge, MA 02139, United States
The Boulevard, Langford Lane, Kidlington, Oxford OX5 1GB, United Kingdom

Library of Congress Cataloging-in-Publication Data
A catalog record for this book is available from the Library of Congress

British Library Cataloguing-in-Publication Data
A catalogue record for this book is available from the British Library

ISBN: 978-0-12-802727-1

For information on all Academic Press publications
visit our website at https://www.elsevier.com/

Publisher: Nikki Levy
Acquisition Editor: Rob Sykes
Editorial Project Manager: Karen Miller
Production Project Manager: Julia Haynes
Designer: Matthew Limbert

Typeset by Thomson Digital

Contents

Introduction

When we envisioned writing this book we knew that we wanted to write a handbook and not a textbook. In our minds a textbook is a book that evenhandedly covers the advantages and disadvantages of the subject under discussion and does not excessively impose the authors points of view or experience into the subject matter. An example would be Lawless and Heymann (2010). On the other hand we view a handbook as a practical manual drawing on the authors' expertise to guide the reader to the desired endpoint. In this case that endpoint is deriving information for business and scientific reasons from sensory and instrumental data collected on alcoholic beverages. In order to obtain information, the reader needs to understand which tests should be performed and why these should be chosen rather than others. Then once the data have been collected the appropriate statistical technique(s) should be used to analyze the data and lastly the data should be interpreted in the light of general and specific knowledge of the particular product and questions asked in the study.

In this book we are drawing on our three decades of experiences in the "trenches" of sensory and instrumental analyses and we hope to provide the reader, whether a novice in these fields or not, with guide posts to help with the collection, analysis, and interpretation of sensory and instrumental data related to alcoholic beverages. We discuss a variety of alcoholic beverages, but many of our examples come from the wine world and can usually be readily applied to other types of alcoholic beverages. We feel that this book is ideally suited for readers new to one or both of the areas discussed in the book, or it could also be used as a textbook for a beginning class in chemical analyses or sensory analyses of alcoholic beverages.

The book includes four general topic areas and different readers may find use in one or more of these areas. The first area covered in Chapter 1 reviews sensory perception specifically of the visual, olfactory, gustatory, and in-mouth tactile senses. The overarching narrative in this chapter is about natural human variation in the senses. We do not all see, smell, taste, and feel alcoholic beverages the same way—we each live in our own sensory world and the only way to obtain interpretable data in the face of this human diversity is through the use of sensory panels rather than individuals. The second topical area of the book describes the uses of sensory testing in the life cycle of alcoholic beverages from the beginning as raw material through marketing and drivers of consumer liking. In this topical area, Chapter 3 covers the sensory tests that could be used to answer different questions. We cover analytical sensory tests, such as difference testing, descriptive analysis, temporal dominance of sensation, time–intensity testing, as well as consumer preference and hedonic tests. For each of these we suggest the most appropriate candidates and discuss the data analyses associated with each test.

The next topical area includes Chapters 4, 5, and 6 and covers chemical analysis of alcoholic beverages. We first review the general purposes of chemical testing. As in the sensory analysis chapter, we provide an overview of the uses of chemical analyses from the raw materials to final products. We also discuss the development of a process analytical chart that analysts can use to ensure that appropriate analyses are performed throughout the beverage production process. Next we cover the classical, rapid, "wet chemistry" methods that are commonly performed in alcoholic beverage analysis laboratories, including measurement of pH, titratable acidity, soluble solids, and ethanol. UV-visible spectroscopic and enzymatic methods, as well as newer indirect methods using infrared spectroscopy and sensor-based approaches are also noted. In the final chapter of this section we discuss gas and liquid chromatography, mass spectrometry, capillary electrophoresis, and NMR methods for analysis of volatile and nonvolatile components. We provide a brief background of the theory of these instrumental methods as well as selected applications to alcoholic beverages. Not all analytical laboratories will have access to these instruments, however the methods and information obtained are often critical for relating the composition of alcoholic beverages to their sensory properties.

Finally, the book contains 10 case studies. The goal of these are to allow the novice reader to take a data set and completely analyze it using R as the data analysis program. Each case study has an objective and then shows the reader how to attain that objective. The following areas are covered in the case studies:

1. Difference testing, including a priori and posthoc power calculations. The data set is based on Chardonnay wine.
2. Descriptive analysis and data analyses using multivariate analysis of variance (MANOVA), univariate analyses of variance (ANOVAs), and canonical variate analyses (CVA). Again the data set is based on Chardonnay wines, in this case from Australia.
3. Another descriptive analysis case where we again analyze sensory descriptive data MANOVA and ANOVAs but we then use principal component analysis (PCA) as the multivariate data analysis technique. The data set is based on USA mead wines.
4. In this case study we use sparkling wines as the data set and this allows us to discuss the factors to consider when doing sensory analysis with a product containing dissolved carbon dioxide and to introduce the use of analysis of covariance (ANCOVA).
5. Here we look at astringency data over time using time–intensity analyses. This allows us to discuss the issues associated with palate cleansing and the effects of salivary flow rates on perceived astringency. The data set is based on a red wine.

6. In this case study we discuss the issues associated with temporal dominance of sensation (TDS) data. We used TDS to characterize the taste and mouthfeel TDS curves for three very different wines, a semisweet Riesling, a dry Rosé, and a Grenache–Syrah–Mourvedre Blend.
7. We now turn our attention to a rapid sensory method, sorting which allows us a holistic overview of the difference among a set on cream liqueurs. We analyze the data using multidimensional scaling (MDS).
8. Next we discuss another rapid sensory technique projective mapping or napping to characterize the holistic differences in the aromas of a set of gins from the USA, England, France, and Germany. We show how one can use multifactor analysis (MFA) to analyze the data.
9. In this case study we were interested in evaluating consumer hedonic data for California Cabernet Sauvignon wines using ANOVA and internal preference mapping. Additionally, we used descriptive analysis (DA) data to interpret consumer liking data using external preference mapping.
10. In the last case study we were interested in determining how changes in chemical properties during storage at different temperatures change the sensory attributes of a wine. Additionally, we wanted to determine whether the closure type changed these effects. We again used MFA to analyze these data, providing an example of statistical analysis of data from two very different platforms, that is, chemical/volatile profiles and sensory profiles.

As a group these case studies and the attendant R-codes should allow the novice reader to collect, analyze, and interpret most of the data that they would need for a wide variety of applications.

We would like to sincerely thank our alpha reader, Helene Hopfer for all the hours that she spent reading our drafts, discussing sensory and chemical analyses, as well as data analysis issues. Any errors that remain are despite her best efforts to the contrary. We would also like to thank our postdoctoral fellows, graduate students, and undergraduate students for their inspiration, questions, and passion. Last but not least we would like to thank our husbands, Bill and John, for their endless patience and support.

Reference

Lawless, H.T., Heymann, H., 2010. Sensory Evaluation of Food: Principles and Practices, second ed. Springer Science + Business Media, New York, NY, USA.

In this chapter

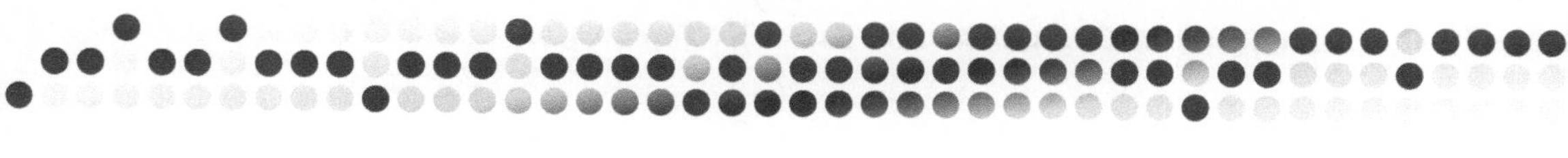

Sensory perception

"Flavor is the psychological interpretation of the physiological response to a physical stimulus." This definition, said to one of us in 1986, by Milton Bailey, a professor of meat flavor at the University of Missouri, has resonated with us ever since. It neatly encapsulates the three entities required to perceive flavor—a human with a functional brain, a working sensory system, and a physical stimulus in the form of chemical compounds. This chapter focuses on the human sensory system and the transduction of information to the brain. We will briefly describe how the sensory systems for vision, olfaction, gustation and chemesthesis work. We will also describe some of the related neurobiology research relevant to each of these systems. Finally, we will highlight the variability in these systems from person to person.

1.1 Vision

Humans see because the photoreceptors in the human eye respond to the visible spectrum (about 380–760 nm) of the electromagnetic spectrum which encompasses radio waves to cosmic rays (MacKinney and Little, 1962). When light waves strike an object, a portion of those waves reflect off the object. The light that is reflected from the object is seen by the eye. In other words, a red rose looks red because all the light waves in the visible spectrum except red are absorbed by the rose petals and only red is seen by the eye. Light penetrates the eye through the cornea, pupil, and lens and is focused on the retina in the area of the fovea (Fig. 1.1). In this area there is a mosaic of rods and cones that contains the light-sensitive receptors, also known as opsins or pigments, each pigment responds to light waves of differing lengths. There are about 120 million rods, located to the periphery of the fovea, and these receptors are very sensitive under very low–light conditions and are responsible for scotopic or dark-adapted vision (Kang et al., 2009). The photoreceptor in the rods is rhodopsin and it has a maximal absorption at about 498 nm (Wissinger and Sharpe, 1998). At scotopic levels of illumination the amount of light is below the threshold of the color-sensitive cones but the very sensitive black and white sensing rods are active (Barbur and Stockman, 2010). However, the rods bleach under bright lights and thus would

Sensory and Instrumental Evaluation of Alcoholic Beverages. http://dx.doi.org/10.1016/B978-0-12-802727-1.00001-6

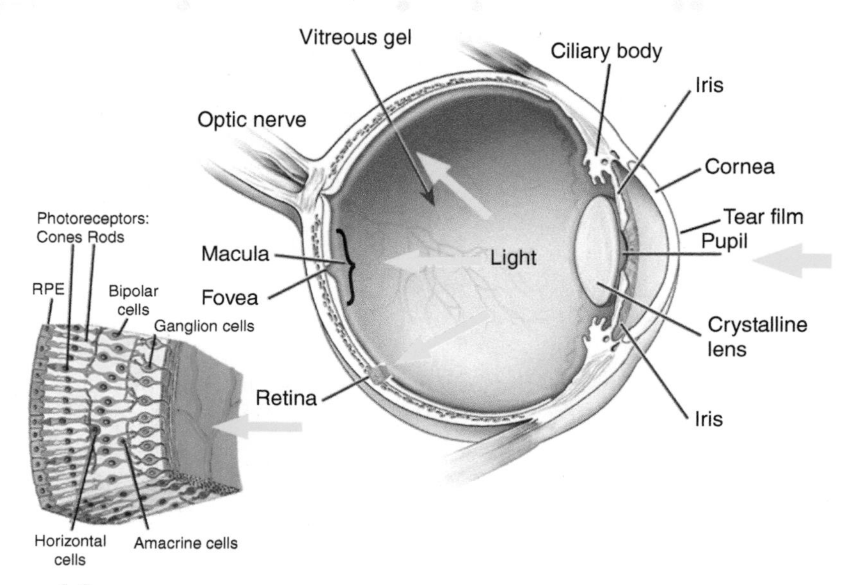

Figure 1.1
Gross anatomy of the human eye and detail of the retina
The major refractive elements and hence primary sources of aberration are the tear film–cornea interface and the crystalline lens. The incident light on the retina is absorbed by the cone and rod photoreceptors after traversing several retinal layers. *Modified from the National Eye Institute, National Institutes of Health (Fuensanta, A.V.-D., Doble, N., 2012. The human eye and adaptive optics. In: Tyson, B. (Ed.), Topics in Adaptive Optics. InTech. Available from: http://www.intechopen.com/books/topics-in-adaptive-optics/the-need-for-adaptive-optics-in-the-human-eye*

not be a factor when we do sensory evaluation of alcoholic beverages. Under mesopic conditions, both rods and the color-sensitive cones are active. At these light levels we are above the threshold of the cones and below the rod saturation level. Mesopic conditions occur under many outdoor nighttime conditions and are especially important to vision in nighttime traffic lighting situations (Kang et al., 2009). Again these would not be the environments used to do sensory evaluation.

The settings used for sensory evaluation would be under photopic conditions where the light receptors in the cones are active, thus we would be above the rod saturation or bleaching levels. There are about 5 million cones located close to the fovea, the area with the highest visual acuity (Sung and Chuang, 2010). Each cone contains one of three different photoreceptors, the long wave or L-opsin, the medium wave or M-opsin, and the short wave or S-opsin. The L-cones are most sensitive to light at 558 nm or red colors, with the M-cones most sensitive at about 530 nm or green colors (Wissinger and Sharpe, 1998). Finally, the S-cones are most sensitive at about 426 nm or blue colors (Wissinger and Sharpe, 1998).

The pigments in the cones are typical G-protein coupled receptors (GPCRs) with seven transmembrane α-helices and a light-absorbing chromophore, 11-*cis*-retinal, is found covalently bonded to the receptor pigment (Shichida and Imai, 1998). The chromophore is a vital component of the process of phototransduction by which light energy in the form of photons is converted to a signal is then sent to the brain via the optic nerve. Specifically, during the visual cycle a photon is biologically converted into an electrical signal in the retina (Fig. 1.2). The light photon isomerizes the 11-*cis*-retinal to all transretinal which in turn activates the G-protein transducin by exchanging guanosine 5′-diphosphate) for guanosine 5″-triphosphate (Sung and Chuang, 2010). This dissociates the α-subunit of transducin which activates cyclic guanosine 3′,5′-monophosphate (cGMP)-phosphodiesterase which hydrolyzes cytoplasmic cGMP. The decreased concentration of cGMP causes the cGMP-gated cation channels on the plasma

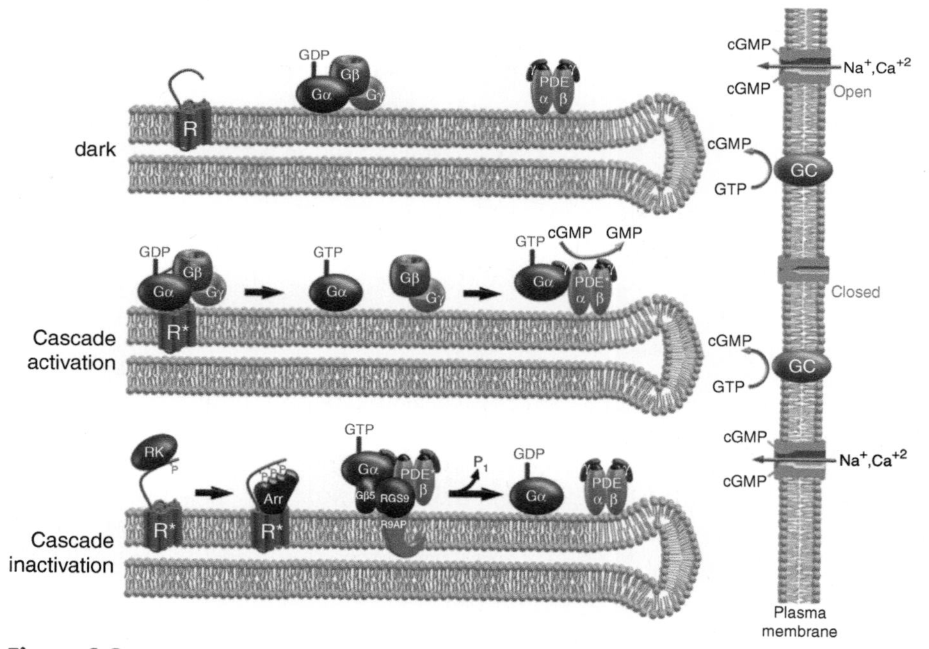

Figure 1.2

Schematic of the phototransduction cascade activation and inactivation

The upper disc illustrates inactive cone opsin *(R)*, transducin (Gα, Gβ, and Gγ subunits), and phosphodiesterase *(PDE)* specifically α, β, and γ subunits in the dark. The reactions in the middle disc illustrate light-induced transducin and PDE activation. The reactions in the lower disc represent R* inactivation via phosphorylation by cone opsin kinase *(RK)* followed by arrestin *(Arr)* binding and transducin/PDE inactivation by the RGS9-1·Gβ5·R9AP complex. Guanylate cyclase *(GC)* is shown as a component of the plasma membrane; in real cells it is likely to be present within the disc membranes as well.

Reprinted with permission from Burns, M.E., Arshavsky, V.Y., 2005. Beyond counting photons: trails and trends in vertebrate visual transduction. Neuron 48, 387–401.

membrane to close. This leads to electrotonic conduction through the bipolar cells to the ganglion cells which are the beginning of the optical nerve (Reese, 2011).

The ganglion cells project to the lateral geniculate nucleus (LGN) in the thalamus. The LGN is the nucleus that acts as a connector between the retina and the visual cortex (Conway et al., 2010). The visual cortex (V1 or the striate cortex) receives the largest projection from the LGN and it in turn feeds the higher visual areas (V2, V3, and V4). While the V1 cortex has been extensively studied, there is still controversy as to how much computation V1 does (Ng et al., 2007), but this controversy and the specific cell types in the V1 and other visual cortexes are outside the parameters of this chapter. Please see McKeefry et al. (2009), Ng et al. (2007), and Grill-Spector and Malach (2004) for more information.

The three different light receptors make normal human color vision trichromatic. However, approximately 3–8% of males and less than 0.5% of females are "color blind." In most cases these individuals lack one or more fully functional L- or M-cone receptor. There are different types of color blindness depending on which of the L- or M-cones are absent or not fully functional (Table 1.1). According to

Table 1.1 Color blindness and variation in humans

Types of variation	Expressed phenotype	Incidence	Pigments found in cones[a]	Cytogenetic location on specified chromosome
Protanopia	Red blind	1.01% males, 0.02% females	S and M	Xq28
Protanomaly	Red weak	1.08% males, 0.03% females	S, M, and M-like	Xq28
Deuteranopia	Green blind	1% males, 0.1% females	S and L	Xq28
Deuteranomaly	Green weak	5% males, 0.35% females	S, L, and L-like	Xq28
Tritanopia	Blue–yellow blind	<1:1,000		7q32
Tritanomaly	Blue–yellow weak	<1:1,000		7q32
Blue cone monochromasy	Severe lack of color discrimination with other visual disorders	<1:100,000		Xq28
Achromatopsia or rod monochromasy	Complete lack of color vision with severe photophobia	<1:30,000		2q11, 8q21, 1p13

[a]*Keep in mind that individuals with normal vision are trichromats and have L-, M-, and S-pigments.*

Source: Based on Deeb, S.S., 2004. Molecular genetics of color-vision deficiencies. Visual Neurosci. 21, 191–196; Sharpe, L.T., Stockman, A., Jägle, H., Nathans, J., 2001. Opsin genes, cone photopigments, color vision and color blindness. In: Gegenfurtner, K.R., Sharpe, L.T. (Eds.), Color Vision from Genes to Perception. Cambridge University Press, pp. 3–52; Neitz, J., Neitz, M., 2011. The genetics of normal and defective color vision. Vision Res. 51, 633–651.

Wissinger and Sharpe (1998) these types are true dichromats who would either by "red blind" (protanopes) or "green blind" (deuteranopes). Protanopes have dichromatic vision and only have M- and S-cones, similarly deutaranomalous individuals also have dichromatic vision and only have L- and S-cones (Deeb, 2004). There are also individuals that are "red weak" (protanomalous) or "green weak" (deuteranomalous). Protanomalous individuals have S-, M-, and anomalous M-like pigment. In these cases the L-opsin had been mutated to be more similar to an M-opsin. On the other hand, deuteranomalous individuals have S-, L-, and an anomalous L-like pigment. For these individuals their M-opsin had been mutated to be more similar to an L-opsin. In both cases this is due to unequal crossover mutations between the female X chromosomes during gamete formation (Deeb, 2004). These deficiencies can be traced to the red/green photoreceptor gene on chromosome Xq28 (Deeb and Motulsky, 1996). Color blindness due to S-cones deficiencies are rarer. Tritanopia and tritanomaly are characterized by a confusion of blue–yellow and occur in less than 1 in 1000 cases. This color blindness is due to autosomal mutations on chromosome 7 (Deeb, 2004). Blue cone monochromasy is extremely rare (less than 1 in 100,000) and it is an X chromosome–linked condition with severe lack of color discrimination, as well as other visual-related disorders (Deeb, 2004). Finally, some individuals have achromatopsia or total color blindness. These individuals (less than 1 in 30,000) have rod monochromasy and have severe photophobia and no color perception (Deeb, 2004). This type of color blindness occurs due to mutations on chromosome 2 (Wissinger and Sharpe, 1998).

In addition to the variants of color blindness described earlier, there are also major variations in the L- and M-pigments of humans with normal trichromatic color vision. In color vision testing an anomaloscope is often used (Rüfer et al., 2012). In this test the individual is asked to make a metameric match by using two different light sources to match a specified color. In the first of these, the Rayleigh match, the subject is asked to use red light and green light to match yellow light. This allows the investigator to test red/green color deficiencies. In the Moreland match the subject uses indigo light and green light to match a cyan light (Deeb, 2004).

We know from the Rayleigh match that individuals with normal color vision will use different percentages of red and green colors to match the yellow color. This indicates that they "see" the red and/or green colors differently. DNA testing allows us to determine the genetic differences in the genes expressing the opsins and relating those results to the outcomes of the Rayleigh test. There are at least two reasons for these variations in normal color vision, first the number of opsin gene arrays on the X chromosome and second, single nucleotide polymorphisms (Sharpe et al., 1998). In the first instance each pigment array gene has one red pigment gene and one to five green pigment genes. In the second instance the polymorphisms lead to the situation where at a specific location on the expressed protein a single amino acid is substituted by another.

The most common situation is the substitution of serine for alanine at the 180 location of the L- or the M-pigment, with about 60% of individuals encoding for serine and about 40% for alanine. In the case of the red or L-pigment this substitution can lead to a red shift in the wave length of maximum sensitivity of about 4 nm when serine is present instead of alanine, in other words these individuals need less red light in the red/green light blend to match the yellow (Sharpe et al., 1998). Similarly, about 90% of individuals encode for alanine on the green or M-pigment and about 10% for serine.

As we have seen in this section all humans do not see colors in the same way. If color perception is important in the specific sensory evaluation of alcoholic beverages we need to keep in mind both the presence of color blindness and of the variations in normal color perception. In these cases we would probably need to screen out color blind individuals (Thiadens et al., 2013). This is relatively simple to do by asking the potential sensory panelists to perform a color blindness evaluation using a pseudoisochromatic test, such as the Ishihara test or an arrangement test, such as the Farnsworth D-15 test. If the potential panelist is color blind and if the study is on the color of a beverage they should be eliminated from the panel. However, it is not easy to test for variations in color vision with potential panelists with normal trichromatic vision. This is one of the reasons that sensory scientists use relatively large panels to "average" out the natural variability in the human senses rather than using single or a few individuals.

1.2 Gustation

In general in every day conversation humans tend to use the word "taste" when they actually mean smell, for example, one may say that the ice cream "tastes" like strawberries when in actual fact the ice cream smells like strawberries. The sense of taste seems to be quite simple with humans possessing taste receptor cells that allow them to perceive five primary tastes, namely, sweet, sour, salty, bitter, and umami. Umami is a Japanese word that translates as brothy or savory (Stillman, 2002). The tastants usually associated with these tastes are sucrose (sweet); tartaric, malic, or citric acids (sour); sodium chloride (NaCl) (salty); caffeine or quinine (bitter); and monosodium glutamate (MSG) (umami). There is some speculation that humans may have taste receptor cells sensitive to fatty acids (Tucker et al., 2014) but this has not been shown conclusively, and since alcoholic beverages (with the exception of cream liquors) usually do not contain fats this is of less relevance to this chapter. Humans also perceive sensations, such as astringency, burn or irritation, tingling, and temperature in the mouth but these sensations are not mediated by the taste buds and are thus not part of the sense of taste. We will discuss these mouthfeel sensations later in the chapter when we describe the common chemical sense or chemesthesia.

The taste receptor cells are located in onion-shaped taste buds (Fig. 1.3) that are located in papillae scattered on the tongue. The circumvallate or "wall-like"

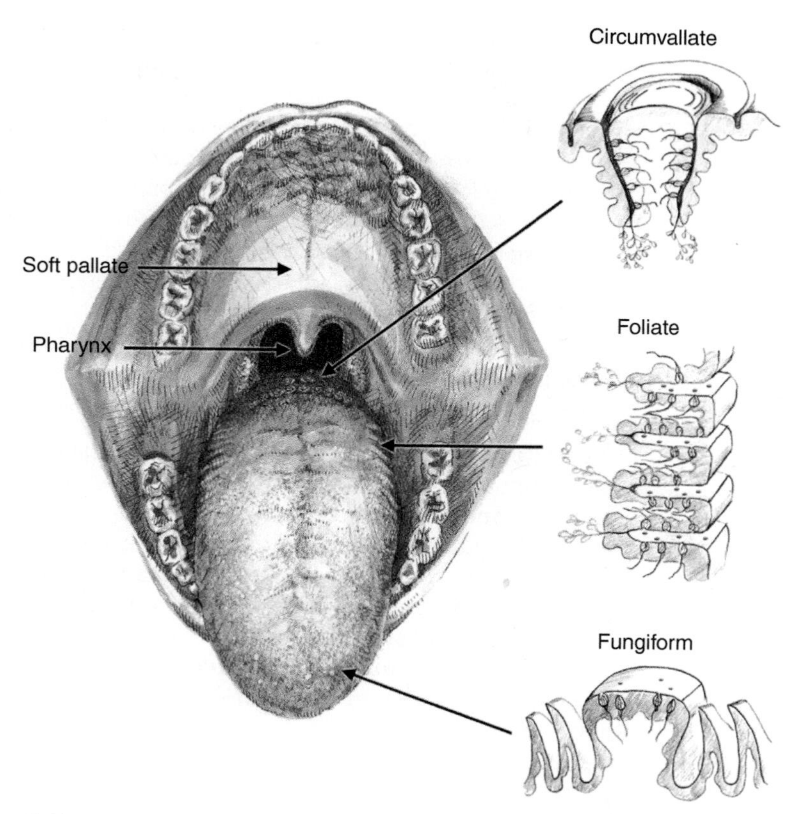

Figure 1.3
This diagram shows the locations of the papillae containing taste buds on the human tongue
The anterior tongue contains taste buds within fungiform (mushroom shaped) papillae innervated by the chorda tympani branch of cranial nerve VII (facial). Posterior tongue contains taste buds within the trenches of foliate (leafy) and circumvallate (walled tower shaped) papillae innervated by the lingual-tonsilar branch of CN IX (glossopharyngeal). The glands appearing below foliate and circumvallate papillae (insets) are Von Ebner's glands that secrete into the folds of the papillae. Soft palate contains taste buds on the surface of the epithelial sheet without papillary structures innervated by the greater superficial petrosal branch of the CN VII. The taste buds posterior to the pharynx are innervated by the superior laryngeal branch of CN X (vagus). *Reprinted with permission from Breslin, P.A.S., Spector, A.C., 2008. Mammalian taste perception. Curr. Biol. 18, R148–R155.*

papillae, containing hundreds of taste buds, are found toward the back of the tongue where about 12 of these papillae are arranged in a chevron pattern pointing toward the throat (Stillman, 2002). These papillae are innervated by the glossopharyngeal nerve as are the foliate papillae. The foliate or "leaf-like" papillae, with hundreds of taste buds, are located on each side of the back-third of the tongue. The fungiform or "mushroom-like" papillae, each with only a few taste buds, are scattered across the front two-thirds of the surface of the tongue where they are innervated by the chorda tympani branch of the facial nerve (Kinnamon, 2012). Humans also have a fourth type of papillae, the filiform or "thread-like" papillae that do not carry taste buds but these papillae are involved with the tactile (touch) sensation on the tongue (Smith and Margolskee, 2001). There are also functional taste buds in the soft palate of the human throat that are not associated with papillae and these taste buds are innervated by the greater superficial petrosal nerve, a branch of the lingual nerve (Stillman, 2002). The taste buds on the epiglottis and the larynx are innervated by the superior laryngeal nerve, a branch of the vagus nerve (Kinnamon, 2012). This abundance of taste buds and nerves innervating the various lingual and oral locations makes it very difficult for someone to lose their sense of taste, as we will see later, the opposite is true for the sense of smell (Breslin and Spector, 2008).

Each taste bud contains 50–100 cells, some basal cells and support cells, as well as taste receptor cells (Kinnamon, 2012). Taste buds are continuously being regenerated and the average life span of a taste bud cell is about 10 days (Breslin and Spector, 2008). The taste receptor cells have microvilli (finger-like projections) that push through the opening, also known as the taste pore, at the top of the taste bud. Each taste receptor cell is sensitive to one of the five basic tastes. This means that all tastes can be perceived across the tongue in all locations and that the so called "tongue map" does not exist (Bartoshuk, 1993). There are documented cases where individuals have lost the front two-thirds or their whole tongue but their sense of taste was essentially unaffected due to the taste buds found in the soft palate (Bartoshuk, 1993). The tastants, chemicals that have been dissolved either in the saliva or that are ingested in liquid form, come in contact with the microvilli and the interactions lead to electrical changes in the taste cells that prompt the cells to send chemical signals to the brain (Smith and Margolskee, 2001). The different tastants have different transduction pathways in their specific taste receptor cells. For sweet, bitter, and umami tastants the taste receptor cells have GPCRs but for the sour and salty tastants the transduction occurs through ion channels (Kinnamon, 2012).

For sweet tastants the chemicals do not enter the cell but they bind to the dimeric GPCR composed of T1R2 and T1R3. These receptors bind to sugars, synthetic sweeteners, D-amino acids, and some sweet proteins (Kinnamon, 2012). For umami tastants the dimeric receptor contains T1R1 and T1R3 GPCRs and in humans the T1R1 + T1R3 dimeric GRPCR binds to L-glutamate and the binding is enhanced by the presence of 5′-ribonucleotides, such as inosine-5-monophosphate and guanosine-5-monophosphate. In the case of bitter tastants, the tastants bind to a family of about 25–30 T2R GPCRs. Some of the T2R GPCRs are broadly tuned and respond to many bitter tastants, for example, *TAS2R14* [Following standard nomenclature a human gene name is abbreviated as all italicized capitals (*TAS1R1*) while rat/mouse genes are abbreviated as italicized capital and smaller case letters (*Tas1r1*). Additionally, the corresponding protein symbol would be T1R1 (Bachmanov and Beauchamp, 2007)] which responds to at least 68 compounds (Roland et al., 2013). This may explain why humans do not easily distinguish bitter tastes from different chemical sources (Kinnamon, 2012). On the other hand there are T2R GPCRs that are quite narrowly tuned to one or a few specific bitter tastants, for example, D-amygdalin is one of the few tastants causing a response in *TAS2R16* (Meyerhof et al., 2010). For further information please see Wiener et al. (2012). Once the tastants bind to the T1R2 + T1R3 GPCR, in the case of sweet substances, or the T1R1 + T1R2 GPCR in the case of L-glutamate, or the T2R family of GPCRs for bitter substances the subsequent signaling process is very similar.

The binding of the taste receptor splits a heterotrimeric G-protein that usually consists of Gα-gustducin (α-gust) and its $\beta\gamma$ cohort ($\beta3\gamma13$) into α and $\beta\gamma$ subunits and this activates a phospholipase (Cβ2) enzyme. This then activates a series of second messengers which releases calcium (Ca^{++}) ions in the cell which triggers an electrochemical charge (or depolarization) which in turn triggers the release of neurotransmitters (chemical signals). The nerve cells (neurons) receive these signals and send them to the brain (Kinnamon, 2012; Chaudari and Roper, 2010; Smith and Margolskee, 2001).

In the case of salts, such as NaCl the tastants trigger the receptor cells when the sodium ions (Na^+) enter the cell through the microvilli (or through the side surfaces of the cell) (Smith and Margolskee, 2001). The concentration of Na^+ in the cell increases and this triggers depolarization that results in Ca^{++} entering the cell. The Ca^{++} triggers the release of neurotransmitters which the neurons receive and this sends a signal to the brain. The ion channel receptor cell types for salt perception is different from the ion channel receptor cells for acid as shown by the Zuker laboratory (Huang et al., 2006; Chandrasekhar et al., 2010).

Acid tastants (sour compounds) generate hydrogen (H^+) ions that have three ways of interacting with the taste receptor cell; by blocking potassium (K^+) ion channels on the microvilli; by directly entering the cell; and by binding to and opening the ion channels on the microvilli that allow positive ions to enter the cell. In all cases the cells experiences an increase in positive ion concentrations which trigger a depolarization and the release of neurotransmitters (Smith and Margolskee, 2001).

The neurons transmit the signals from the GPCRs to the brain stem, specifically the nucleus of the solitary tract of the medulla, and then to the gustatory cortex in the parvicellular part of the ventroposterior medial nucleus of the thalamus (VPMpc) (de Araujo and Simon, 2009). The gustatory cortex receives its signals from the VPMpc. A discussion whether the spatial coding in the cortex is based on the "labeled-line" theory or the "across-neuron" pattern of both is beyond the scope of this book but reviews by Lemon and Katz (2007), de Araujo and Simon (2009), and Di Lorenzo (2000) sheds some light on the controversy. The Zuker laboratory (Chen et al., 2011) also published a gustotopic map of the mammalian brain that shed additional light on the situation.

Similarly to vision, there are wide ranges of sensitivity to specific tastants due to human genetic diversity. For example, some humans are not sensitive to low concentrations of quinine (a bitter tastant) while others are. This result has been traced to a few specific loci on chromosome 12 (Ledda et al., 2014). It also seems that two different taste receptors mediate low and high concentrations of quinine, respectively (Ledda et al., 2014). There also seems to be a variation in the perception of coffee bitterness due to taste receptors genes associated with chromosome 7 (Hayes et al., 2013) but this genetic variation did not predict coffee intake or liking. Although it has not been extensively studied it does seem that there are also polymorphisms with the sweet receptors genes that lead to differential thresholds for sweet tastants and possibly also to differential pleasantness ratings for above threshold concentrations of sugars (Garcia-Bialo et al., 2009; Hayes et al., 2013). Based on human genomic studies comparing African, Asian, European, and Native American DNA sequences there are distinct variations in the genes associated with T1R1 and T1R3, which would suggest variability in sensitivity to umami tastants (Garcia-Bialo et al., 2009). Additionally, a group of European adults indicated that about 27% were unable to distinguish 29 mM isomolar concentrations of MSG and NaCl, but in this case the individuals were not genetically characterized (Lugaz et al., 2002). The genetic variability to sour and salty tastants has not been extensively studied, however, there are indications that these also show diversity, see for example, Wise et al. (2007) and Hayes et al. (2013).

Unlike the other taste modalities, the variability in taste sensitivities has been extensively studied for bitter perception because humans have widely divergent detection thresholds (the concentration at which a stimulus is first detected) to two compounds, phenylthiocarbamide (PTC) and to 6-n-propylthiouracil (PROP). Both these compounds are thioureas and have a thiocyanate moiety (N—C=S). Neither of these compounds have a food-related origin but solutions of these compounds stimulate a bitter to intensely bitter sensation in the majority of individuals (the "medium tasters" and "supertasters," respectively) while a minority (the "nontasters") do not find these solutions bitter. There is larger variation in the proportion of "nontasters": about 30% of Caucasian North Americans, about 40% of individuals from the Indian subcontinent, between 6 and 23% of Chinese, and about 3% of West Africans (Garcia-Bialo et al., 2009). PTC sensitivity has been traced to specific haplotypes of the *TAS2R38* gene where position 49 is either encoded for proline (P) or alanine (A), position 262 is either A or valine (V) and position 296 is either encoded for V or isoleucine (I) (Bufe et al., 2005). The most frequent haplotypes are PAV and AVI with AAI, AAV, and PVI occurring less commonly. Individuals that were homologous to PAV are the most responsive to PTC at threshold concentrations and they are usually associated with "supertasters," whereas homologous AVI individuals are not responsive to PTC and they are "nontasters." AAI, AAV, and PVI are associated with intermediate tasters at threshold levels but are more similar to PAV/PAV homologous individuals at above threshold concentrations. It should be noted that AVI/AVI individuals usually do not have a general taste insensitivity to other bitter and sweet compounds but that they are specifically insensitive to PTC. The *TAS2R38* haplotypes are less predictive for PROP bitterness than PTC bitterness sensitivity indicating that additional genetic controls are operational for PROP (Bufe et al., 2005).

There has been some indications that individuals responsive to PTC/PROP tend to dislike cruciferous vegetables and alcohol and that young female "supertasters" tend to dislike tart citrus, coffee, sweet, and fatty foods (Drewnowski et al., 1999; Duffy and Bartoshuk, 2000). However, for Brazilian older woman there were very modest effects of polymorphisms in *TAS2R38* and food consumption in the food groups of interest (Colares-Bento et al., 2012). Yet, Öter et al. (2011) showed that PROP-insensitive school children were significantly more likely to have a higher caries risk than PROP-sensitive individuals. Additionally, Carrai et al. (2011) found that in Polish and Czech populations PROP nontasters had an increased risk of colorectal cancer compared to the taster group.

Today, there is a growing body of evidence that there are individuals with heightened taste sensitivities that may or may not be related to PROP sensitivity (Hayes and Keast, 2011). Thus these authors suggest that distinguishing between "genetic supertasters," namely, PAV/PAV homozygotes (pST) that may or may not be PROP sensitive or "generalized supertasters" that have intense responses to sucrose and quinine but not PROP (gST) may lead to "an alphabet soup of prefixes (pST, gST, uSt, sST, etc.)." Instead these authors suggest that a new term is needed to describe "broad, elevated chemosensory response across (taste) stimuli" and their candidate term is hypergeusia.

These variations in taste sensitivities have an effect on the sensory evaluation of alcoholic beverages since alcohol itself is perceptibly much more bitter and more irritating (see in Section 1.4 for why) to "tasters" than "nontasters" (Bachmanov et al., 2003). It also seems clear that PROP "tasters" tend to consume less alcohol than "nontasters" (Hayes et al., 2013). Pickering et al. (2013) have also shown that there is a modest relationship between wine experts and PROP responsiveness (with wine experts more likely to be super or medium tasters than nontasters) but that this relationship did not occur for *foodies* (individuals with high involvement with and liking of food). If required for the specific sensory project it is relatively simple to broadly classify nontasters and tasters by using filter paper disks impregnated with 50 mmol/L PROP (available from MP Biomedicals, Ohio, USA, http://www.mpbio.com) and a 100 mm generalized labeled magnitude scale with the following cut offs: nontasters <9 mm, medium tasters 9–50 mm, and supertasters >50 mm (Bartoshuk et al., 2004; Pickering et al., 2013). However, there is some recent evidence that supertasting is not necessarily related to papillae density (Gameau et al., 2014) and thus one should proceed with some caution.

1.3 Olfaction

As we had seen earlier, gustation or the sense of taste is relatively simple, with only five (or possibly six) modalities in humans. On the other hand the sense of smell is much more complex. First, humans can probably differentiate among about one trillion odors (Bushdid et al., 2014) and second, the same olfactory system is used as two distinct senses, based on orthonasal and retronasal olfaction, respectively (Auvray and Spence, 2008). If the stimulus is food-related, retronasal olfaction leads to neural processing across the cortex, while for nonfood stimuli and for orthonasal olfaction a much more localized processing occurs in the cortex (Table 1.2).

The nose has two nasal passages separated by the nasal septum and each nasal passage has an external opening (naris or nostril). The olfactory receptor (OR) cells are located in the olfactory epithelium below the cribriform plate (part of the skull). In order to be perceived, a volatile compound must enter the nasal cavity from the nostril (orthonasallly) or from the throat (retronasally) and be absorbed into the mucus associated with the olfactory epithelium (Doty and Kamath, 2014). The olfactory neurons are highly ciliated and the cilia form a "carpet" that interact with the volatiles absorbed into the olfactory mucus. The olfactory neurons project directly through the cribriform plate into the olfactory bulb in brain (Shepherd, 2006). Information is then transmitted from the olfactory

Table 1.2 The dual olfactory system

Operations	Orthonasal olfaction	Retronasal olfaction
Stimulation route	Through the external nares	From the back of the mouth through the nasopharynx
Stimuli	Floral scents, perfumes, smoke, food aromas, pry/predator smells, social odors, pheromones, MHC molecules	Food volatiles
Processed by	Olfactory pathway influenced by the visual pathway	Olfactory pathway combined with pathways for taste, touch, sound, and active sensing by proprioception to form a "flavor system"
Brain systems involved	OE, OB, LOFC, MOFC, AM, OC	OE, OB, LOFC, MOFC, AM, OC, LH, insula taste, DI cortex, AVI, ACC, motor cortex, SOM taste, SOM faces, SOM tongue, PPC, VPM thalamic nucleus, LH, VC, NST, taste buds, and cranial nerves V, VII, IX, and X

ACC, accumbens; AM, amygdala; AVI, anterior ventral insular cortex; DI, dorsal insular; LH, lateral hypothalamus; LOFC, lateral orbitofrontal cortex; MHC, major histocompatibility complex molecules (MHC Sequencing Consortium, 1999. Complete sequence and gene map of a human major histocompatibility complex. Nature 401, 921–923); MOFC, medial orbitofrontal cortex; NST, nucleus of the solitary tract; OB, olfactory bulb; OC, olfactory cortex; OE, olfactory epitheliu; PPC, posterior parietal cortex; SOM, somatosensory; VC, visual cortex; VPM, ventral posteriomedial.

Based on Table 1 and Fig. 2 and used with permission from Shepherd, G.M., 2006. Smell images and the flavor system in the human brain. Nature 444, 316–321.

bulb to the olfactory cortex where patterns of neuronal signals are transformed into odor images (Shepherd, 2006; Gottfried, 2010; Weiss and Sobel, 2012).

The OR proteins are expressed in the cilia and each olfactory cell has only one type of OR expressed. There are over 350 functional ORs and the subgenome of the ORs constitutes about 1% of the human genome (Auffarth, 2013). It seems as if the "olfactory code is made up of activated sets of overlapping receptor cells that can be viewed as spatial maps within both the epithelium and the olfactory bulb" (Doty and Kamath, 2014). The ORs are G-coupled proteins, part of the same families seen in vision and gustation, and during signal transduction the odorant molecule binds to the OR which leads to an activated receptor (R*) (Mainland et al., 2014). The R* couples to the olfactory G-protein (G_{olf}) and this increases the production of cyclic 3′-5′-cyclic adenosine monophosphate (cAMP). The cAMP opens the cyclic nucleotide channels into the cell that allows Ca^{++} ions to enter the cell. Simultaneously, a channel mediating the efflux of Cl^{-} ions occurs. This results in a transduction current that is passed to the olfactory neuron cell body where it initiates a train of action potential that are sent to the glomeruli in the olfactory bulb. See Fig. 1.4 for further transduction.

The air flow through the two sides of the nose tends to vary due to swelling of the veins in the nasal turbinate bones. The nasal cycle varies from 25 to 200 min and is idiosyncratic to the individual (Baranuik and Kim, 2007). The consequence of this is that the air flow from the two nares combined remains constant while the air flow in one is faster and in the other is slower. This allows hydrophilic odorants to absorb better onto the ORs in a high flow rate situation and conversely allow for the better absorption of hydrophobic odorants at low flow rates (Sobel et al., 1999). Additionally, the differential flow rate allows us to localize the direct of an odorant although it seems that this is mostly due to trigeminal rather than OR stimulation (Frasnelli et al., 2009).

There are codes for about 850 ORs in humans but due to mutations creating pseudogenes only about 350 are expressed. "Variation between humans is such that, statistically, it is unlikely that any two humans use the same set of ORs" (Sell, 2014). This leads to the concept of specific anosmia where an individual with an otherwise normally functioning olfactory system either is unable to smell as specific odorant or has a higher than average threshold for that odorant. Since many odorants also have trigeminal effects in the nose the method of determining specific anosmia can lead to widely divergent rates, for example, the specific anosmia for androstenone (associated with boar taint) ranges from 1.8 to 75% of the population, depending on the specific experimental protocol (Sell, 2014). Odorants associated with specific anosmia and that may be relevant in an alcoholic beverage context are acetic acid, dimethyl sulfide, isobutyraldehyde, diacetyl, isoamyl alcohol, isobutyric acid, dimethyl disulfide, isovaleric acid, methional, phenylethanol, amyl acetate, isoamyl acetate, cinnamaldehyde, geranial, vanillin, geraniol, 1,8-cineole, eugenol, geosmin, β-damascenone, β-ionone, farnesol, and others (Sell, 2014). However, specific anosmia does not only occur due to

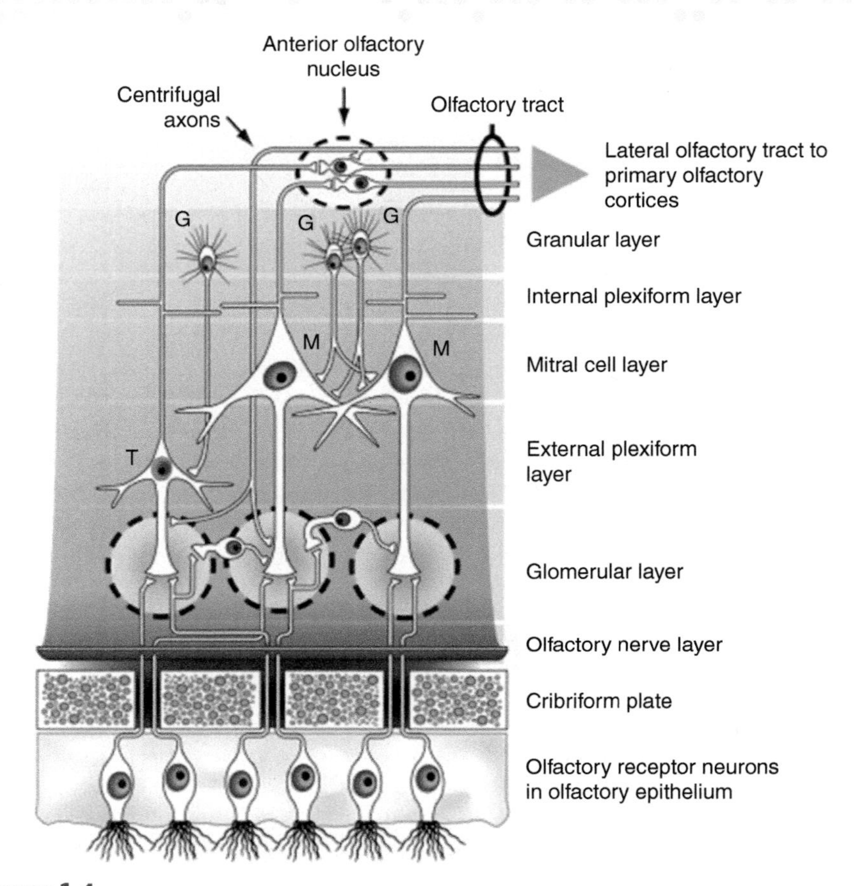

Figure 1.4
Olfactory bulb neuroanatomy schematic
Bipolar olfactory receptor (OR) neurons of the olfactory epithelium project axons through the cribriform plate that diverge upon entering the olfactory bulb to form the olfactory nerve layer and coalesce with axons from OR neurons with identical OR genes to form glomeruli. In the glomeruli, OR neurons synapse with mitral *(M)* and tufted *(T)* cells and are modulated by periglomerular neurons and centrifugal axons from brainstem nuclei. Granule *(G)* cells project from the granular to modulate M and T cells. M and T cells are the main output neurons of the olfactory bulb, projecting to the intrabulbar portion of the anterior olfactory nucleus, as well as the primary olfactory cortices via the lateral olfactory tract. *Used with permission from Duda, J.E., 2010. Olfactory systems pathology as a model of Lewy neurodegenerative disease. J. Neurol. Sci. 289, 49–54.*

differential expression of ORs since it has been shown that for some but not all odorants, an individual may experience specific anosmia but once they are exposed to the odorant overtime they will become able to smell the odorant (Tempere et al., 2012). This could occur due to a learning effect (a name for a pattern associated with the odorant) or there may be a hormonal or direct effect on the expression of latent ORs in the olfactory epithelium (Sell, 2014). The consequence of the variation in specific anosmia across individuals is that each of us probably live in our own unique sensory universe. Once again this would lead us to do sensory evaluation studies with groups of individuals rather than relying on the expertise (and fallibility) of a single or few experts.

1.4 Chemesthesia

The mucosa of the eye, the nose, and the mouth are all innervated by branches of the trigeminal nerve (cranial nerve V) and in the mouth and the nose, the chemosensory perception by the trigeminal nociceptive neurons add to the concept of flavor perception (Auvray and Spence, 2008). Signals from these neurons activate cortical areas associated with irritation and pain, with responses encompassing pain, irritation, sneezing, vasodilation, tearing, nasal secretion, sweating, and bronchoconstriction (Patel and Pinto, 2014). In the nose, irritants, such as ammonia, ethanol, other alcohols, acetic acid, carbon dioxide, capsaicin, and menthol all stimulate the nociceptors associated with the trigeminal nerve. In the mouth there is a wide range of somatosensory systems ranging from the trigeminal nociceptors where many of the nasal irritants also act as oral irritants, to tactile sensations, such as astringency associated with plant flavonoid phenols (tannins).

A number of transient receptor potential (TRP) neurons associated with the lingual trigeminal nerve have been identified. Specifically, TRPV1 is the chemoreceptor for capsaicin (the "hot" compound found in chili peppers), TRPM8 is the chemoreceptor for cooling compounds, such as menthol, eucalyptol, linalool, and geraniol among others while TRPA1 responds to the pungency associated with cinnamon and thiol reactive compounds, such as allyl isothiocyanate in mustard and diallyl disulfide in garlic (Damann et al., 2008). Of interest to those studying alcoholic beverages, the TRPV1 receptor is also activated by ethanol and this stimulation may cause the burning or irritant sensations associated with oral ethanol (Brasser et al., 2014). Additionally, TRPV1 also responds to acidic pH, in the wine range, and temperatures above 43°C (Damann et al., 2008).

Tannins are broadly found in many common beverages, such as tea, coffee, wine, and beer. Astringency is a mouth drying/puckering sensation associated with a loss of lubrication and possibly also the precipitation of salivary proteins (Schwarz and Hofmann, 2008). Intriguingly, Schöbel et al. (2014) conclusively showed that astringency is a trigeminal sensation involving the activation of G protein–coupled signaling. They found that strongly activating compounds had to contain one or more galloyl moieties and that activation responses depended on Ca^{++} influx and CA^{++} but not on TRPs. Variations in perceived astringency between individuals are related to differences in salivary flow rates and/or salivary protein content. Individuals with low salivary flow tend to rate wines as more astringent than people with high salivary flow rates (Ishikawa and Noble, 1995; Fischer and

Noble, 1994). Dinnella et al. (2011) found that high responder groups (based on their salivary protein content) reported apple, carrot, and grape juice spiked with tannic acid to be more astringent, compared to low responder groups. However, some researchers doubt that there is a direct relationship between the perception of astringency and precipitation of salivary proteins (Schwarz and Hofmann, 2008). Finally, Furlan et al. (2014) showed that tannins disordered the hydrophobic interior of lipid vesicles and also created a new lipid–tannin topology in lipid droplets. They speculate that during wine tasting tannins can precipitate lipids thus diminishing the tannin concentration available to precipitate mucosal proteins.

1.5 Flavor

In the previous sections we have described the senses associated with the evaluation of flavor but the perception of flavor occurs due to "the integration of information provided by multiple unisensory inputs: olfactory, gustatory, somatosensory, auditory, visual, and trigeminal information are all combined" (Shankar et al., 2010). The result for a sensory scientist is that we always need to keep in mind that the senses affect each other. In many ways therefore, flavor is a perceptual modality occurring due to integration of information from all the senses (see Auvray and Spence, 2008 for an excellent review).

There are numerous studies showing that the color of a beverage affects the perceived sweetness (see Auvray and Spence, 2008) and perceived odor of the beverage (see Spence et al., 2010, for a review). Additionally, humans have a tendency to orally refer retronasal odors to the mouth rather than the ORs in the nose (Spence, 2013) and this is why we often say that we have lost our sense of "taste" when we have a cold, when in reality we have temporarily lost our sense of smell due to congestion in the nose. Similarly color affects odor, as Gottfried and Dolan (2003) stated in there review "the nose smells what the eye sees." For example, Morrot et al. (2001) showed that a white wine colored red with flavorless red food coloring would elicit red-related flavor descriptors, such as raspberry or cherry from wine experts. The same wine, not colored red, did not elicit these types of red fruit–related odor words. This result is not surprising since the color of the beverage led these experts the wrong direction. In a preliminary study these authors had shown, through triangle discrimination testing in opaque glasses, that the red-colored and original white wines were indistinguishable. Thus if the wine experts had evaluated the red-colored wine in black glasses or with blindfolds, the lack of seeing the wine color would not have led them astray.

Additionally, it is not only the integration of the various senses that affect the perceptual modality of flavor. In a series of studies Spence and coworkers have shown the effects of context, receptacles, cutlery, ambient noise, ambient light, and so on on the perception of food and beverage flavors. A few examples will be briefly described. They found that coffee consumed from white mugs were rated to be less sweet than coffee consumed from transparent or blue mugs (Van Doorn et al., 2014). Similarly the color of the plate used to serve a dessert affect the consumers' perception of the food but the effect was varied with the type of dessert (Piqueras-Fiszman et al., 2013). Yogurts were perceived to be denser when the spoon used to sample the yogurt was a lighter plastic spoon rather than a heavier weight plastic spoon (Harrar and Spence, 2013). Spence et al. (2014) speculate that the only taste unaffected by noise is umami and that this is why many passengers order umami-rich tomato juice or Bloddty Mary drinks on airplanes. Finally, Velasco et al. (2013) showed that room ambiance due to audiovisual displays affected the participants perception of odors (woody and grassy), as well as tastes.

In conclusion, everything discussed in this chapter once again reminds us, that when we do sensory evaluation of alcoholic beverages we are using humans as our measuring devices and that humans a variable in their genetic and phenotypic capabilities, that humans are affected by the multisensory integration of their senses and last but not least that humans are affected by their multisensory environment. As sensory scientists we need to make sure that our studies are designed to take these facts into account and to produce valid, repeatable results. In Chapter 3 we will be discussing sensory methods and how to stabilize and/or circumvent some of these sources of variability.

References

Auffarth, B., 2013. Understanding smell—the olfactory stimulus problem. Neurosci. Biobehav. Rev. 37, 1667–1679.

Auvray, M., Spence, C., 2008. The multisensory perception of flavor. Conscious. Cogn. 17, 1016–1031.

Bachmanov, A.A., Beauchamp, G.K., 2007. Taste receptor genes. Annu. Rev. Nutr. 27, 389–414.

Bachmanov, A.A., Kiefer, S.W., Molina, J.C., Tordoff, M.G., Duffy, V.B., Bartoshuk, L.M., et al., 2003. Chemosensory factors influencing alcohol perception, preferences and consumption. Alcohol. Clin. Exp. Res. 27, 220–231.

Baranuik, J.N., Kim, D., 2007. Nasonasal reflexes, the nasal cycle and the sneeze. Curr. Allergy Asthma Rep. 7, 105–111.

Barbur, J.L., Stockman, A., 2010. Photopic, mesopic and scotopic vision and changes in visual performance. Besharse, J., Dana, R., Dartt, D.A. (Eds.), Encyclopedia of the Eye, vol. 3, Elsevier, Oxford, UK, pp. 323–331.

Bartoshuk, L.M., 1993. Genetic and pathological taste variation: what can we learn from animal models and human disease? In: Chadwick, D., Marsh, J., Goode, J. (Eds.), The Molecular Basis of Smell and Taste Transduction. John Wiley and Sons, Chichester, England, pp. 251–267.

Bartoshuk, L.M., Duffy, V.B., Green, B.G., Hoffman, H.W., Ko, C.W., Lucchina, L.A., et al., 2004. Valid across-group comparisons with labeled scales: the gLMS versus magnitude matching. Physiol. Behav. 82, 109–114.

Brasser, S.M., Castro, N., Feretic, B., 2014. Alcohol sensory processing and its relevance for ingestion. Physiol. Behav. 2.

Breslin, P.A.S., Spector, A.C., 2008. Mammalian taste perception. Curr. Biol. 18, R148–R155.

Bufe, B., Breslin, P.A.S., Kuhn, C., Reed, D.R., Tharp, C.D., Slack, J.P., et al., 2005. The molecular basis of individual differences in phenylthiocarbamide and propylthiouracil bitterness perception. Curr. Biol. 15, 322–327.

Bushdid, C., Magnasco, M.O., Vosshall, L.B., Keller, A., 2014. Humans can discriminate more than 1 trillion olfactory stimuli. Science 343, 1370–1372.

Carrai, M., Steinke, V., Vodicka, P., Pardini, B., Rahner, N., Holinski-Feder, E., et al., 2011. Association between *TAS2R38* gene polymorphisms and colorectal cancer risk: a case-control study in two independent populations of Caucasians origin. PLoS One 6, e20464, pp. 1–7.

Chandrasekhar, J., Kühn, C., Oka, Y., Yarmalinsky, D.A., Hummler, E., Ryba, N.J.P., et al., 2010. The cells and peripheral representation of sodium taste in mice. Nature 464, 297–302.

Chaudari, N., Roper, S.D., 2010. The cell biology of taste. J. Cell Biol. 190, 285–296.

Chen, X., Gabitto, M., Peng, Y., Ryba, N.J.P., Zuker, C.S., 2011. A gustotopic map of taste qualities in the mammalian brain. Science 333, 1262–1266.

Colares-Bento, F.C.J., Souza, V.C., Toledo, J.O., Moraes, C.F., Alho, C.S., Lima, R.M., et al., 2012. Implication of the G145C polymorphism (rs713598) of the *TAS2R38* gene on food consumption by Brazilian older women. Arch. Gerontol. Geriatr. 54, e13–e18.

Conway, B.R., Chatterjee, S., Field, G.D., Horwitz, G.D., Johnson, E.N., Koida, K., et al., 2010. Advances in color science: from retina to behavior. J. Neurosci. 30, 14955–14963.

Damann, N., Voets, T., Nilius, B., 2008. TRPs in our senses. Curr. Biol. 18, R880–R889.

De Araujo, I.E., Simon, S.A., 2009. The gustatory cortex and multisensory integration. Int. J. Obes. 33, S34–S43.

Deeb, S.S., 2004. Molecular genetics of color-vision deficiencies. Vis. Neurosci. 21, 191–196.

Deeb, S.S., Motulsky, A.G., 1996. Molecular genetics of human color vision. Behav. Genet. 26, 195–207.

Di Lorenzo, P.M., 2000. The neural code for taste in the brain stem: response profiles. Physiol. Behav. 69, 87–96.

Dinnella, C., Recchia, A., Tuorila, H., Monteleone, E., 2011. Individual astringency responsiveness affects the acceptance of phenol-rich foods. Appetite 56, 633–642.

Doty, R.L., Kamath, V., 2014. The influences of age on olfaction: a review. Front. Psychol. 5, 1–20.

Drewnowski, A., Henderson, S.A., Levine, A., Hann, C.S., 1999. Taste and food preferences as predictors of dietary practices in young women. Public Health Nutri. 2, 513–519.

Duffy, V.B., Bartoshuk, L.M., 2000. Food acceptance and genetic variation in taste. J. Am. Diet. Assoc. 100, 647–655.

Fischer, U., Noble, A.C., 1994. The effect of ethanol, catechin concentration and pH on the sourness and bitterness of wine. Am. J. Enol. Vitic. 45, 6–10.

Frasnelli, J., Charbonneau, G., Collignon, O., Lepore, F., 2009. Odor localization and sniffing. Chem. Senses 34, 139–144.

Furlan, A.L., Castets, A., Nallet, F., Pianet, I., Grélard, A., Dufourc, E.J., Géan, J., 2014. Red wine tannins fluidify and precipitate lipid liposomes and bicelles. A role for lipids in wine tasting? Langmuir 30, 5518–5526.

Gameau, N.L., Nuessle, T.M., Sloan, M.M., Santorico, S.A., Coughlin, B.C., Hayes, J., 2014. Crowdsourcing taste research: genetic and phenotypic predictors of bitter taste perception as a model. Front. Integr. Neurosci. 8, 1–8.

Garcia-Bialo, B., Toguri, C., Eny, K.M., El-Sohemy, A., 2009. Genetic variation in taste and its influence on food selection. OMICS 13, 69–80.

Gottfried, J.A., Dolan, R.J., 2003. The nose smells what the eye sees: crossmodal visual facilitation of human olfactory perception. Neuron 39, 375–386.

Gottfried, J.A., 2010. Central mechanisms of odour perception. Nat. Rev. Neurosci 11, 828-664.

Grill-Spector, K., Malach, R., 2004. The human's visual cortex. Annu. Rev. Neurosci. 27, 649–6477.

Harrar, V., Spence, C., 2013. The taste of cutlery: how the taste of food is affected by the weight, size, shape, and color of the cutlery used to eat it. Flavour 2, 21.

Hayes, J.E., Keast, R.S.J., 2011. Two decades of supertasting: where do we stand? Physiol. Behav. 104, 1072–1075.

Hayes, J.E., Feeney, E.L., Allen, A.L., 2013. Do polymorphisms in chemosensory genes matter for human ingestive behavior? Food Qual. Prefer. 30, 202–216.

Huang, A.L., Chen, X., Hoon, M.A., Chandrasekhar, J., Guo, W., Tränker, D., et al., 2006. The cells and logic for mammalian sour taste detection. Nature 442, 934–938.

Ishikawa, T., Noble, A.C., 1995. Temporal perception of astringency and sweetness in red wine. Food Qual. Prefer. 6, 27–33.

Kang, I., Reem, R.E., Kaczmarowski, A.L., Malpeli, J.G., 2009. Contrast sensitivity of cats and humans in scotopic and mesopic conditions. J. Neurophysiol. 102, 831–840.

Kinnamon, S.C., 2012. Taste receptor signaling—from tongues to lungs. Acta Physiol. 204, 158–168.

Ledda, M., Kutalik, Z., Souza Destito, M.C., Cirillo, C.A., Zamboni, A., Martin, N., et al., 2014. GWAS of human bitter taste perception identifies new loci and reveals additional complexity of bitter taste genetics. Hum. Mol. Genet. 23, 259–267.

Lemon, C.H., Katz, D.B., 2007. The neural processing of taste. BMC Neurosci. 8 (Suppl. 5), 1–8.

Lugaz, O., Pillias, A.M., Faurion, A., 2002. A new specific ageusia: some humans cannot taste L-glutamate. Chem. Senses 27, 105–115.

MacKinney, G., Little, A.C., 1962. Color of Foods. The AVI Publishing Company, Inc., Westport, Connecticut.

Mainland, J.D., Lundström, J., Reisert, J., Lowe, G., 2014. From molecule to mind: an integrative perspective on odor intensity. Trends Neurosci. 37, 443–454.

McKeefry, D.J., Gouws, A., Burton, M.P., Morland, A.B., 2009. The noninvasive dissection of the human visual cortex: using fMRI and TMS to study the organization of the visual brain. Neuroscientist 15, 489–506.

Meyerhof, W., Batram, C., Kuhn, C., Brockhof, A., Chudoba, E., Bufe, B., Appendino, G., Behrens, M., 2010. The molecular receptive ranges of human TAS2R bitter taste receptors. Chem. Senses 35, 157–170.

Morrot, G., Brochet, F., Dubourdieu, D., 2001. The color of odors. Brain Lang. 79, 309–320.

Ng, J., Bharath, A.A., Zhaoping, L., 2007. A survey of the architecture and function of the primary visual cortex (V1). EURASIP J. Adv. Signal Process. 2007, 17.

Öter, B., Ulukapi, I., Ulukapi, H., Topçuog˘lu, N., Çildir, ., 2011. The relation between 6-n-propylthiouracil sensitivity and caries activity in school children. Caries Res. 45, 556–560.

Patel, R.M., Pinto, J.M., 2014. Olfaction: anatomy, physiology and disease. Clin. Anat. 27, 54–60.

Pickering, G.J., Jain, A.K., Bezawada, R., 2013. Super-tasting gastronomes? Taste phenotype characterization of *foodies* and *wine experts*. Food Qual. Prefer. 28, 85–91.

Piqueras-Fiszman, B., Giboreau, A., Spence, C., 2013. Assessing the influence of the color of the plate on the perception of a complex food in a restaurant setting. Flavour 2, 24–35.

Reese, B.E., 2011. Development of the retina and optic pathway. Vision Res. 51, 613–632.

Roland, W.S.U., van Buren, L., Gruppen, H., Driesse, M., Gouka, R.J., Smit, G., Vincken, J.-P., 2013. Bitter taste receptor activation by flavonoids and isoflavonoids: modeled structural requirements for activation of hTAS2R14 and hTAS2R39. J. Agric. Food Chem. 61, 10454–10466.

Rüfer, F., Sauter, B., Klaussner, A., Göbel, K., Flammer, J., Erb, C., 2012. Age-corrected reference values for the Heidelberg multi-color anomaloscope. Graefes Arch. Clin. Exp. Ophthalmol. 250, 1267–1273.

Schöbel, N., Radtke, D., Kyereme, J., Wollmann, N., Cichy, A., Obst, K., et al., 2014. Astringency is a trigeminal sensation that involves the activation of G protein-coupled signaling by phenolic compounds. Chem. Senses 39, 471–487.

Schwarz, B., Hofmann, T., 2008. Is there a direct relationship between oral astringency and human salivary protein binding? Eur. Food Res. Technol 227, 1693–1698.

Sell, C.S., 2014. The mechanism of smell. Chemistry and The Sense of Smell. Wiley, New York, pp. 32–187.

Shankar, M.U., Levitan, C.A., Spence, C., 2010. Grape expectations: the role of cognitive influences in color-flavor interactions. Conscious. Cogn. 19, 380–390.
Sharpe, L.T., Stockman, A., Jägle, H., Knau, H., Klausen, G., Reitne, A., et al., 1998. Red, green and red-green pigments in the human retina: correlations between deduced protein sequences and psychophysically measured spectral sensitivities. J. Neurosci. 18, 10053–10069.
Shepherd, G.M., 2006. Smell images and the flavor system in the human brain. Nature 444, 316–321.
Shichida, Y., Imai, H., 1998. Visual pigment: G-protein-coupled receptor for light signals. Cell. Mol. Life Sci. 54, 1299–1315.
Smith, D.V., Margolskee, R.F., 2001. Making sense of taste. Sci. Am. 284, 32–39.
Sobel, N., Khan, R.M., Saltman, A., Sullivan, E.V., Gabrieli, J.D.E., 1999. The world smells different to each nostril. Nature 402, 35.
Spence, C., 2013. Multisensory flavor perception. Curr. Biol. 23, R365–R369.
Spence, C., Levitan, C.A., Shankar, M.U., Zampini, M., 2010. Does food color influence taste and flavor perception in humans. Chem. Percept. 3, 68–84.
Spence, C., Michel, C., Smith, B., 2014. Airplane noise and the taste of umami. Flavour 3, 2.
Stillman, J.A., 2002. Gustation: intersensory experience par excellence. Perception 31, 1491–1500.
Sung, C.-H., Chuang, J.-Z., 2010. The cell biology of vision. J. Cell Biol. 190, 953–963.
Tempere, S., Cuzange, E., Bougeant, J.C., de Revel, G., Sicard, G., 2012. Explicit sensory training improves the olfactory sensitivity of wine experts. Chemosens. Percept. 5, 205–213.
Thiadens, A.A.H.J., Hoyng, C.B., Polling, J.R., Bernaerts-Biskop, R., van den Born, L.I., Klaver, C.C.W., 2013. Accuracy of four commonly used color vision tests in the identification of cone disorders. Ophthalmic Epidemiol. 20, 114–122.
Tucker, R.M., Mattes, R.D., Running, C.A., 2014. Mechanisms and effects of "fat taste" in humans. BioFactors 40, 313–326.
Van Doorn, G.H., Wuillemin, D., Spence, C., 2014. Does the color of the mug influence the taste of the coffee? Flavour 3, 10.
Velasco, C., Jones, R., King, S., Spence, C., 2013. Assessing the influence of the multisensory environment on the whisky drinking experience. Flavour 2, 23.
Weiss, T., Sobel, N., 2012. What's primary about primary olfactory cortex. Nat. Neurosci 15, 10–12.
Wiener, A., Shudler, M., Levit, A., Niv, M.Y., 2012. BitterDB: a database of bitter compounds. Nucleic Acids Res. 40, D413–D419.
Wise, P.M., Hansen, J.L., Reed, D.R., Breslin, P.A., 2007. Twin studies on the heritability of recognition thresholds for sour and salty taste. Chem. Senses 32, 39–57.
Wissinger, B., Sharpe, L.T., 1998. New aspects of an old theme: the genetic basis of human color vision. Am. J. Hum. Genet. 63, 1257–1262.

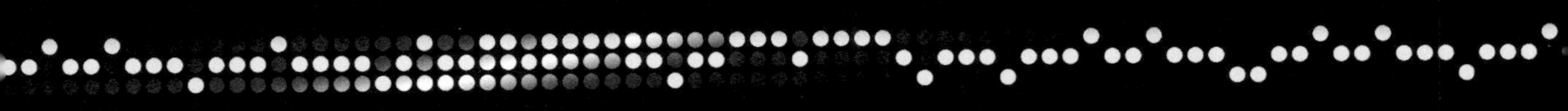

In this chapter

Uses and good practices of sensory evaluation in the alcoholic beverage industry

2.1 Introduction

Before we describe the uses of sensory evaluation in the analysis of alcoholic beverages we need to define sensory evaluation. In 1975, the Sensory Evaluation Division of the Institute of Food Technologists (IFT) defined sensory evaluation as "a scientific discipline used to evoke, measure, analyze, and interpret those responses to products that are perceived by the senses of sight, smell, touch, taste, and hearing" (IFT, 1975). In other words, sensory scientists use humans and their sensory systems as measuring devices. However, depending on the questions asked of the subjects, there are two types of sensory evaluation, namely, analytical sensory evaluation and hedonic or consumer sensory evaluation (Lawless and Heymann, 2010).

For analytical sensory evaluation the sensory scientist is interested in the product. In this case the sensory methods will allow one to determine whether the product differs from other product(s) (discrimination testing), and if it does differ, how it differs in terms of specific sensory attributes [descriptive analysis (DA)]. It is also possible to determine how the product differs from others holistically (sorting or projective mapping), without breaking product differences down into specific sensory attributes. One can also determine the changes in the product sensory attributes over time as the product is consumed (time intensity testing or temporal dominance of sensations). All these methods in parentheses will be described in Chapter 3. In all cases with analytical sensory testing the main focus is the product, the data collected are objective, and these techniques usually utilize trained panelists.

In consumer sensory science the data collected are subjective in nature since the sensory scientist is interested in the consumers' response to the product. The most used consumer scale is the nine-point hedonic scale which ranges from like extremely through neither like nor dislike to dislike extremely, although as

Sensory and Instrumental Evaluation of Alcoholic Beverages. http://dx.doi.org/10.1016/B978-0-12-802727-1.00002-8

we will see later other scales are also used. Paired preference testing is also used when there are only two products to compare as is preference ranking for three to four products (see Chapter 3). In all cases for consumer sensory evaluation, the sensory scientists would not utilize trained panelists but actual and/or would-be consumers of the specific product.

In the next sections we will briefly describe the uses of sensory evaluation in all phases of the production (through the use of quality assurance programs) and marketing of alcoholic beverages.

2.2 Sensory evaluation in the alcoholic beverage production and marketing cycle

2.2.1 Tracking the sensory effects of changes in the production of source ingredients

As discussed in Chapter 4 we frequently conduct chemical analyses of the source ingredients but it is less common to do a formal sensory evaluation of these ingredients. For example, winemakers use the flavor, including tastes and mouthfeel, of grapes to determine the optimal date for harvesting, however, this is usually one winemaker or two walking through a vineyard tasting the grapes and deciding on the ripeness of the crop, without any formal "training" or assessment. Since the early 2000s, some formal sensory evaluation of grapes has been performed, starting with work by Rousseau and Delteil (2000) and then followed, in Australia, by Winter et al. (2004). This grape berry assessment is based on the use of trained panelists and is a modified version of DA (see Chapter 3). In most situations this work is done in research centers and universities (Olarte Mantilla et al., 2013) rather than in wineries, although there is nothing to prevent wineries from doing so.

For other alcoholic beverages the picture is less clear. In the case of beer, it is highly unusual to do formal sensory evaluation of the barley, water, and malt ingredients, but for spirits, sherries, and sparkling wines, the source product is a base wine, whether from grapes or other fruits, and a formal sensory study of the attributes of the base can lead to information on the potential attributes of the finished product, see for example, Torrens et al. (2010) and De La Presa-Owens et al. (1998). In all of these cases, the techniques used to assess the base wines are also modifications of DA.

2.2.2 Tracking the sensory effects of changes in the production process

Application of sensory evaluation to monitor changes during production is more frequently found than applications described in Section 2.2.1. The production process for alcoholic beverages spans the fruit (in many cases) or grain (in many other cases), the growing conditions and areas, the fermentation process, aging, distillation, infusions, etc.

In the case of spirits there is a reasonable amount of chemical data on the effect of the initial fruit cultivar on the resultant product (S´liwin´ska et al., 2015; Tsakiris et al., 2014) but in general there is a scarcity of sensory data, and often the quality of the sensory data is suspect. An exception is the study by Vázquez-Araújo et al. (2013) on the effects of two hop cultivars on the sensory attributes of herbal liqueurs. They found that one of the cultivars (Nugget) created a liqueur that was too bitter and had too much herbal flavor while the other cultivar (Saaz) increased the intensity of aromatic hop flavors while balancing the perception of sweetness in the liqueur. Also, Husson et al. (2004) studied the effect of apple cultivars on the sensory effects of monovarietal ciders and found that the ciders made from the same cultivar were quite similar but the ciders made from different cultivars could be clearly distinguished sensorially.

In the wine realm, sensory studies of the effect of grape origin abound. For example, Lund et al. (2009), Robinson et al. (2012), and King et al. (2014) all used DA to study the effect of region of origin on the sensory attributes of New Zealand Sauvignon blanc, Australian Cabernet Sauvignon, and Californian and Mendoza Malbec wines, respectively. Bauer et al. (2011) used DA to authenticate German Riesling *terroirs*. Other researchers used DA to study the effect of grape vine water status (Marcinak et al., 2013; Ledderhof et al., 2014), or vine vigor on the sensory attributes of the resultant wines (Bramley et al., 2011). Other environmental effects, such as the effects of smoke from brush and bush fires on the wines made from smoke-tainted grape berries have been extensively studied in recent years, see Krstic et al. (2015).

Once the grapes are in the winery, numerous sensory studies have been conducted on the effects of, for example, cold soaking (Casassa et al., 2015), yeast strain (King et al., 2010, 2011), lactic acid bacteria (López et al., 2011; Delaquis et al., 2000). At the completion of the fermentation phase sensory studies on the effect of extended maceration in the production of red wines (Casassa et al., 2013) are also found. During wine aging the effects of oak barrels and/or oak origin (Francis et al., 1992), barrel alternatives (Ortega-Heras et al., 2010; Gutiérrez Afonso, 2002), microoxidation (Oberholster et al., 2015), as well as blending (Hopfer et al., 2012), and filtering (Buffon et al., 2014) are studied as well. Recently a number of authors have explored the effects of packaging (Hopfer et al., 2012, 2013; Kwiatkowski et al., 2007; Lopes et al., 2009), storage temperatures (Makhotkina et al., 2012; Scrimgeour et al., 2015), and the effect of shipping temperatures on wine sensory attributes (Robinson et al., 2010; du Toit and Piquet, 2014).

The above skims the list of studies exploring the sensory effects of production methods on wine sensory attributes, however, in the nonwine alcoholic beverage field the number of sensory studies exploring the effects of production on sensory attributes is much smaller. It is possible to find some examples for many of the areas described earlier. Sun et al. (2013) used DA to study the effect of coinoculating yeast and lactic acid bacteria on the sensory attributes of cherry wines and found that coculturing increased the fruitiness of the cherry wines. Caldeira et al. (2010) used sensory techniques to study the effects of chestnut and oak barrel alternatives on the sensory attributes of Portuguese brandies. There are also a number of sensory-related articles on the effect of packaging and temperature on the staling of beer, for example, Techakriengkrai et al. (2006). Jack (2003) gives guidelines for the use of DA in the Scotch whisky industry and this may be indicative of future sensory studies on whiskies.

2.3 Consumer sensory evaluation and marketing

2.3.1 The effects of the intrinsic sensory attributes of the beverage on consumer choice and hedonic responses

When choosing to purchase an alcoholic beverage, such as an Irish Cream liqueur for the first time the consumer usually uses a range of extrinsic factors to identify which product to buy. These extrinsic factors are price, label design, label information, country or region of origin, medals, etc. Extrinsic factors usually are outside the scope of the sensory scientist and fall within the remit of market research. However, if the intrinsic factors of the Irish Cream liqueur do not match or exceed the sensory expectations set by the extrinsic factors, then it is very unlikely that the consumer will buy the specific product again. Thus, it is the sensory scientist that has to study the intrinsic attributes of the product and make sure that they match the extrinsic expectations. To do this the products are usually evaluated blind—with no indication of price, label, origin, and so on—and this allows the consumer to only focus on the intrinsic attributes (color, odor, taste, mouthfeel, and aftertaste) to determine their liking of the product and/or their choice for one product over another.

There are numerous consumer studies using a blind tasting protocol in the alcoholic beverage realm. For example, see a wine study by Malherbe et al. (2013) on the effect of *Oenococcus oeni* starter cultures on consumer liking for Pinotage wines. Another example is a beer study by Donadini et al. (2014) on Italian consumers' preferences for bottom-fermented red beers.

2.3.2 The effects of extrinsic factors, such as label, marketing, price, and so on associated with the beverage on consumer choice and hedonic responses

There are also numerous sensory-marketing combination studies linking the effect of extrinsic factors on the consumers' liking for the product. In these studies, the consumers usually evaluate the samples blind (without any extrinsic information) and then they or a different group of consumers do a second evaluation with the extrinsic information.

Lee et al. (2006) evaluated the timing of revealing the extrinsic information on consumers liking responses. They had consumers evaluate two beers (one with a few drops of added balsamic vinegar). The consumer group evaluating the samples with no information had a higher preference (59%) for the adulterated beer than a consumer group told of the adulteration prior to tasting the beers (30%). However, the consumer group told of the adulteration after tasting but before indicating their preference did not significantly decrease their preferences for this beer (52%). In another beer-related study, Guinard et al. (2001) had consumers indicate their liking for 24 beers in a blind condition and then with extrinsic factors present (labels and prices). Through the use of internal preference mapping they found that especially consumers in their 20s significantly changed their hedonic ratings in the second evaluation.

Similarly, in a study of South African wine, Priilaid et al. (2009) found that younger, less-experienced wine drinkers were more likely to change their hedonic scores when given extrinsic information, in this case a quality score by a wine expert. Mueller and Szolnoki (2010) also found a difference between younger, less-experienced consumers and older, more-experienced wine consumers in the impact of extrinsic factors on the liking scores for white wines.

2.4 Integrating analytical sensory data with consumer sensory data with/without additional instrumental and chemical data

2.4.1 Integrating the analytical sensory data with consumer data

Asking consumers directly why they prefer one alcoholic beverage over another or why they like/dislike specific versions of alcoholic drinks is usually very counterproductive. The consumers will list very specific reasons why they like or dislike a specific sample but this information is rarely actionable. Instead, sensory scientists usually determine drivers of liking in a more indirect fashion, through the use of external preference mapping (EPM). As we will see in Chapter 3, in an EPM the consumer data are projected into a two- or three-dimensional space usually created from the DA data, but more rarely, this space may be created by other descriptive methods, such as sorting or napping. The EPM allows the sensory scientist to determine which attributes of the products drive the likes or dislikes of the consumers. These drivers of liking can then be used in the blending, aging, or creation of the products for the market.

There are very few of these studies for nonwine alcoholic beverages. Guinard et al. (2001) studied 24 commercial beers and related sensory descriptive data to consumer data using EPM. Another is a study on hop cultivars for use in herb liqueurs where descriptive data and consumer data were compared using an EPM (Vázquez-Araújo et al., 2013).

There are, however, a number of wine-related EPM studies. Examples are Hopfer and Heymann (2014), Torri et al. (2012), or Frøst and Noble (2002), where drivers of linking and quality perception for specific wines were identified.

2.4.2 Integrating the chemical and instrumental analyses with the analytical sensory data and/or consumer data

It is also possible to use multivariate statistics to create a two- or three-dimensional chemical or instrumental data space that can then be integrated with analytical sensory data (whether derived from DA, sorting, napping, etc.) and/or consumer data. Again, there are not many nonwine studies but a few examples are the Vázquez-Araújo et al. (2013) article described in the previous section where the authors also compared the volatile composition of the two hops cultivars to the analytical and consumer data and found that the herb liqueurs made with the Saaz hops were more preferred than the ones made with Nugget hops. The Nugget hops derived liqueurs were more bitter and had an off-putting "herb flavor." The volatile analyses showed that the Nugget hops liqueur had more aromatic compounds, such as myrcene and humulene. On the wine front, there are a number of such studies, for example, King et al. (2014), who compared sensory descriptive data, chemical data, and climate/topography data for Malbec wines made in Argentina and California and showed regional differences in composition and sensory properties. Additionally, Parpinello et al. (2009) integrated analytical sensory and consumer data with color parameters for Novello red wines, and found that for consumers high-colored wines were associated with high-quality ratings.

2.5 Conclusions

In this chapter we have shown that sensory studies, whether analytical or consumer related are useful throughout the production cycle of alcoholic beverages. We have shown that there are many more wine sensory studies throughout the wine production and marketing cycle than there are nonwine studies. We hope that this book would increase the numbers of nonwine studies in these areas by making the process of doing analytical and consumer sensory studies more accessible.

References

Bauer, A., Wolz, S., Schormann, A., Fischer, U., 2011. Authentication of different terroirs of German riesling applying sensory and flavor analysis. In: Ebeler, S.E., Takeoka, G.R., Winterhalter, P. (Eds.), Progress in Authentication of Food and Wine, vol. 1081, ACS Symposium Series, pp. 131–149, (chapter 9).

Bramley, R.G.V., Ouzman, J., Boss, P.K., 2011. Variation in wine vigour, grape yield and vineyard soils and topography as indicators of variation in the chemical composition of grapes, wine and wine sensory attributes. Aust. J. Grape Wine Res. 17, 217–229.

Buffon, P., Heymann, H., Block, D.E., 2014. Sensory and chemical effects of cross-flow filtration on white and red wines. Am. J. Enol. Vitic. 65, 305–314.

Caldeira, I., Anjos, O., portal, V., Belchior, A.P., Canas, S., 2010. Sensory and chemical modifications of wine-brandy aged with chestnut and oak wood fragments in comparison to wooden barrels. Anal. Chim. Acta 660, 43–52.

Casassa, L.F., Larsen, R.C., Beaver, C.W., Mireles, M.S., Keller, M., Riley, W.R., Smithyman, R., Harbertson, J.F., 2013. Sensory impact of extended maceration and regulated deficit irrigation on Washington State Cabernet Sauvignon wines. Am. J. Enol. Vitic. 64, 505–514.

Casassa, L.F., Bolcato, E.A., Sari, S.E., 2015. Chemical, chromatic and sensory attributes of 6 red wines produced with prefermentative cold soak. Food Chem. 174, 110–118.

De La Presa-Owens, C., Schlich, P., Davies, H.D., Noble, A.C., 1998. Effect of *Méthode Champenoise* process on the aroma of four *V. vinifera* varieties. Am. J. Enol. Vitic. 49, 289–294.

Delaquis, P., Cliff, M., King, M., Girard, B., Hall, J., Reynolds, A., 2000. Effect of two commercial malolactic cultures on the chemical and sensory properties of chancellor wines vinified with different yeasts and fermentation temperatures. Am. J. Enol. Vitic. 51, 42–48.

Donadini, G., Fumi, M.D., Newby-Clark, I.R., 2014. Consumers' preferences and sensory profile of bottom fermented red beers of the Italian market. Food Res. Int. 58, 69–80.

du Toit, W.J., Piquet, C., 2014. Research note: effect of simulated shipping temperatures on the sensory composition of South African Chenin blanc and Sauvignon blanc wines. S. Afr. J. Enol. Vitic. 35, 278–282.

Francis, I.L., Sefton, M.A., Williams, P.J., 1992. A study by sensory descriptive analysis of the effects of oak origin, seasoning, and heating on the aromas of oak model wine extracts. Am. J. Enol. Vitic. 43, 23–30.

Frøst, M.B., Noble, A.C., 2002. Preliminary study of the effect of knowledge and sensory expertise on the liking of red wines. Am. J. Enol. Vitic. 53, 275–284.

Guinard, J.-X., Uotani, B., Schlich, P., 2001. Internal and external mapping of preferences for commercial lager beers: comparison of hedonic ratings by consumers blind versus with knowledge of brand and price. Food Qual. Pref. 12, 243–255.

Gutiérrez Afonso, V.L., 2002. Sensory descriptive analysis between white wines fermented with oak chips and in barrels. J. Food Sci. 67, 2415–2419.

Hopfer, H., Heymann, H., 2014. Judging wine quality: do we need experts, consumers or trained panelists? Food Qual. Pref. 32, 221–233.

Hopfer, H., Ebeler, S.E., Heymann, H., 2012. The combined effects of storage temperature and packaging type on the sensory and chemical properties of Chardonnay. J. Agric. Food Chem. 60, 10743–10754.

Hopfer, H., Buffon, P.A., Ebeler, S.E., Heymann, H., 2013. The combined effects of storage temperature and packaging type on the sensory, chemical and physical properties of a Cabernet Sauvignon wine. J. Agric. Food Chem. 61, 3320–3334.

Husson, F., Bocquet, V., Pagès, J., 2004. Use of confidence ellipses in a PCA applied to sensory analysis application to the comparison of monovarietal ciders. J. Sens. Stud. 19, 510–518.

IFT, 1975. Minutes of Sensory Evaluation Division Business meeting at the 35th Annual Meeting of the Institute of Food Technologists, June 10.

Jack, F., 2003. Development of guidelines for the preparation and handling of sensory samples in the Scotch whiskey industry. J. Inst. Brew. 109, 114–119.

King, E.S., Francis, I.L., Swiegers, J.H., Curtin, C., 2011. Research note: yeast strain-derived sensory differences retained in Sauvignon blanc wines after extended bottle storage. Am. J. Enol. Vitic. 62, 366–370.

King, E.S., Kievit, R.L., Curtin, C., Swiegers, J.H., Pretorius, I.S., 2010. The effect of multiple yeasts co-inoculations on Sauvignon blanc wine aroma composition, sensory properties and consumer preference. Food Chem. 122, 618–626.

King, E.S., Stoumen, M., Buscema, F., Hjelmeland, A.K., Ebeler, S.E., Heymann, H., Boulton, R.B., 2014. Regional sensory and chemical characteristics of Malbec wines from Mendoza and California. Food Chem. 143, 256–267.

Krstic, M.P., Johnson, D.L., Herderich, M.J., 2015. Review of smoke taint in wine: smoke-derived volatile phenols and their glycosidic metabolites in grapes and vines as biomarkers for smoke exposure and their role in the sensory perception of smoke taint. Aust. J. Grape Wine Res. 21, 537–553.

Kwiatkowski, M.J., Skouroumounis, G.K., Lattey, K.A., Waters, E.J., 2007. The impact of closures, including screw cap with three different headspace volumes, on the composition, colour and sensory properties of a Cabernet Sauvignon wine during two years' storage. Aust. J. Grape Wine Res. 13, 81–94.

Lawless, H.T., Heymann, H., 2010. Sensory Evaluation of Foods: Principles and Practices. Springer, New York.

Ledderhof, D., Reynolds, A.G., Manin, L., Brown, R., 2014. Influence of water status on sensory profiles of Ontario Pinot noir wines. LWT—Food Sci. Technol. 57, 65–72.

Lee, L., Frederick, S., Arely, D., 2006. Try it, you'll like it: the influence of expectation, consumption and revelation on preferences for beer. Psychol. Sci. 17, 1054–1058.

Lopes, P., Silva, M.A., Pons, A., Tominaga, T., Lavigne, V., Saucier, C., Darriet, P., Teissedre, P.-L., Dubourdieu, D., 2009. Impact of oxygen dissolved at bottling and transmitted through closures on the composition and sensory properties of a Sauvignon blanc wine during bottle storage. J. Agric. Food Chem. 57, 10261–10270.

López, R., López-Alfaro, I., Gutiérrez, A.R., Tenorio, C., Garijo, P., González-Aranzana, L., Santamariá, P., 2011. Malolactic fermentation of Tempranillo wine: contribution of the lactic acid bacteria inoculation to sensory quality and chemical composition. Int. J. Food Sci. Technol. 46, 2373–2381.

Lund, C.M., Thompson, M.K., Benkwitz, F., Wohler, M.W., Triggs, C.M., Gardner, R., Heymann, H., Nicolau, L., 2009. New Zealand Sauvignon blanc distinct flavor characteristics: sensory, chemical, and consumer aspects. Am. J. Enol. Vitic. 60 (1), 1–12.

Makhotkina, O., Pineau, B., Kilmartin, P.A., 2012. Effect of storage temperature on the chemical composition and sensory profile of Sauvignon blanc wines. Aust. J. Grape Wine Res. 18, 91–99.

Malherbe, S., Menichelli, E., du Toit, M., Tredoux, A., Muller, N., Naes, T., Nieuwoudt, H., 2013. The relationships between consumer liking, sensory and chemical attributes of *Vitis vinifera* L. Cv. Pinotage wines elaborated with different *Oenococcus oeni* starter cultures. J. Sci. Food Agric. 93, 2829–2840.

Marcinak, M., Reynolds, A.G., Brown, R., 2013. Influence of water status on sensory profiles of Ontario Riesling wines. Food Res. Int. 54, 881–891.

Mueller, S., Szolnoki, G., 2010. The relative influence of packaging, labelling, branding and sensory attributes on liking and purchase intent: consumers differ in their responsiveness. Food Qual. Pref. 21, 774–783.

Oberholster, A., Elmendorf, B.L., Lerno, L.A., King, E.S., Heymann, H., Brenneman, C.E., Boulton, R.B., 2015. Barrel maturation, oak alternatives and micro-oxygenation: Influence on red wine aging and quality. Food Chem. 173, 1250–1258.

Olarte Mantilla, S.M., Collins, C., Iland, P.G., Kidman, C.M., Jordans, C., Bastian, S.E.P., 2013. Comparison of sensory attributes of fresh and frozen wine grape berries using Berry Sensory Assessment. Aust. J. Grape Wine Res. 19, 349–357.

Ortega-Heras, M., Pérez-Magariño, S., Cano-Mozo, E., González-San José, M.L., 2010. Differences in the phenolic composition and sensory profile between red wines aged in oak barrels and wines aged with oak chips. LWT—Food Sci. Technol. 43, 1533–1541.

Parpinello, G.P., Versari, A., Chinnici, F., Galassi, S., 2009. Relationship among sensory descriptors, consumer preference and color parameters of Italian Novello red wines. Food Res. Int. 42, 1389–1395.

Priilaid, D., Feinberg, J., Carter, O., Ross, G., 2009. Follow the leader: how expert ratings mediate consumer assessments of hedonic quality. S. Afr. J. Bus. Manage. 40, 51–58.

Robinson, A.L., Adams, D.O., Boss, P.K., Heymann, H., Solomon, P.S., Trengove, R.D., 2012. The influence of geographic origin on the sensory characteristics and wine composition of *Vitis vinifera* cv. Cabernet Sauvignon wines from Australia. Am. J. Enol. Vitic. 63, 467–476.

Robinson, A.L., Mueller, M., Heymann, H., Ebeler, S.E., Boss, P.K., Solomon, P.S., Trengove, R.D., 2010. Effect of simulated shipping conditions on sensory attributes and volatile composition of commercial white and red wines. Am. J. Enol. Vitic. 61, 337–347.

Rousseau, J., Delteil, D., 2000. Presentation d'une method d'analyse sensorielle de baies de raisin. Principe, méthode, interpretation, Revue Française 'Oenologie 183, 10–13.

Scrimgeour, N., Nordestgaard, S., Lloyd, N.D.R., Wilkes, E.N., 2015. Exploring the effect of elevated storage temperature on wine composition. Aust. J. Grape Wine Res. 21, 713–722.

Śliwińska, M., Wiśniewska, P., Dymerski, T., Wardencki, W., Namieśnik, J., 2015. The flavor of fruit spirits and fruit liquors. Flavour Frag. J. 30, 1970207.

Sun, S.Y., Gong, H.S., Zhao, K., Wang, X.L., Wang, X., Zhao, X.H., Yu, B., Wang, H.X., 2013. Co-inoculation of yeast and lactic acid bacteria to improve cherry wines sensory quality. Int. J. Food Sci. Technol. 48, 1783–1790.

Techakriengkrai, I., Paterson, A., Taidi, B., Piggott, J.R., 2006. Staling in two canned lager beers stored at different temperatures—sensory analyses and consumer ranking. J. Inst. Brew. 112, 28–35.

Torrens, J., Riu-Aumatell, M., Vichi, S., López-Tamames, Buxaderas, S., 2010. Assessment of volatile and sensory profiles between base and sparkling wines. J. Agric. Food Chem. 58, 2455–2461.

Torri, L., Noble, A.C., Heymann, H., 2012. Exploring American and Italian consumer preferences for Californian and Italian red wines. J. Sci. Food Agric. 93, 1852–1857.

Tsakiris, A., Kallithraka, S., Kourkoutas, Y., 2014. Grape brandy production, composition and sensory evaluation. J. Sci. Food Agric. 94, 404–414.

Vazquez-Araujo, L., Rodríguez-Solana, R., Cortés-Diéguez, S.M., Domínguez, J.M., 2013. Study of the suitability of two hop cultivars for making herb liqueurs: volatile composition, sensory analysis, and consumer study. Eur. Food Res. Technol. 237, 775–786.

Winter, E., Whiting, J., Rousseau, J., 2004. Wine grape berry sensory assessment in Australia. Winetitles, Adelaide, Australia.

In this chapter

Overview of applicable sensory evaluation techniques

Traditionally sensory methodology has been divided into two groupings, analytical and consumer methods (Lawless and Heymann, 2010). The analytical methods are seen to be objective, frequently use trained judges, and provide information about the product. In contrast, the consumer methods provide subjective information about the consumers' responses to the product and are usually performed by actual consumers of the product category. Some of the rapid sensory techniques do not neatly fall into these groupings and we will highlight these issues when we discuss these methods.

3.1 Discrimination testing

Discrimination testing methods are designed to determine the perceived differences between two samples. The attribute(s) responsible for the perceived difference are not determined in most discrimination tests and these are nondirectional difference tests. The exception to this statement is the use of directional difference tests, such as the 2-alternative forced choice (2AFC also known as paired directional tests) or 3AFC tests. Both tests will allow one to determine the effect of say adding salt to water. In other words, a single ingredient change in concentration leading to change in a single perceptual sensory attribute (e.g., adding salt to water). However, these tests are very rarely useful in the evaluation of alcoholic beverages since changing a single ingredient in these products usually results in changing multiple perceived attributes. Thus, the use of a method that would only provide information about a single perceived attribute change is not useful for more complex samples.

Once the discrimination study is complete the sensory professional will be able to say, with some certainty if the study was designed appropriately, that the sensory perception of the two products did or did not significantly differ. Traditionally, the most frequently used difference tests are the duo–trio and the triangle tests, but recently the tetrad test has also been used more frequently. In this section we

Sensory and Instrumental Evaluation of Alcoholic Beverages. http://dx.doi.org/10.1016/B978-0-12-802727-1.00003-X

will briefly discuss these tests but for a more extensive treatment of these tests the reader is referred to Lawless (2013a,b, Chapters 4 and 5) and Lawless and Heymann (2010).

Duo–trio tests: in the duo–trio test the panelist is provided with three samples, two coded with three-digit random codes and one labeled, reference or R (please note the R has no subscript). The panelist identifies the sample that matches the reference. The underlying concept is that if the panelists can perceive the difference between the two products, then they will correctly match the reference. However, if they cannot perceive the difference then they will randomly pick one of the samples as matching the reference and they would have a one in two chance of guessing correctly. From the sensory professional's perspective there are two types of duo–trio tests, the balanced reference test and the single reference test, respectively. In the balanced reference test both products (A, B) are used as the reference sample and therefore there are four potential serving orders ($R_{(A)}AB$, $R_{(A)}BA$, $R_{(B)}AB$, $R_{(B)}BA$). In the single reference test only one of the products, say A, is used as a reference and thus there are two serving orders ($R_{(A)}AB$, $R_{(A)}BA$). The preferred method is the balanced reference test but in some circumstances, for example, when one product has limited availability or where the panelists are extremely familiar with one of the products the single reference test can be used. When setting up the test one should try to counterbalance the serving orders across the panelists. For example, if there are 50 panelists and the test requires the balanced reference test, approximately 12–13 of each of the 4 serving orders should be used.

Triangle tests: in the triangle test, the panelist is provided with three samples, made up of two different products (i.e., one of the products is presented twice). All three samples are coded with three-digit random codes. The panelist is asked to identify the odd sample. Again, if the panelists can perceive the difference between the two products, then they will correctly identify the odd sample. However, if they cannot perceive the difference then they will randomly pick one of the samples as being odd and they would have a one in three chance of guessing correctly. From the sensory professional's perspective for the two products (A, B) there are six potential serving orders (AAB, ABA, BAA, BBA, BAB, and ABB). Once again, one should counterbalance the serving orders across the panelists.

Tetrad tests: in the tetrad test, the panelist is provided with four samples (i.e., two pairs of each of the two products), all four coded with three-digit random codes. The panelist is asked to group the samples into two groups of two samples each. Again, if the panelists can perceive the difference between the two products, then they will correctly identify the identical samples and create the two groups. However, if they cannot perceive the difference then they will randomly pair the samples and they would have a one in three chance of guessing correctly. From the sensory professional's perspective, for the two products (A, B) there are six potential serving orders (orders AABB, ABBA, BBAA, BAAB, ABAB, and BABA) and one should counterbalance the serving orders across the panelists.

3.1.1 Data analyses for difference tests

The data resulting from difference tests are traditionally analyzed using tables based on binomial expansion (the so-called Roessler tables), by using z-tests or by using χ^2-tests. More recently, Thurstonian scaling has become an additional technique used to analyze these data(Lawless, 2013a). In this section, we will briefly describe some of these methods, as well as the importance of statistical power in interpreting the results of difference tests. In the case studies, we will describe the use of an R-package (sensR created by Christensen and Brockhoff, 2015) that will allow us to analyze discrimination tests using the R-platform.

All the difference tests are one-tailed tests because in all cases the sensory scientist knows which answer from the panelist is the correct answer. As we will see further in this chapter the paired preference test (a consumer related test) is two-tailed.

Binomial expansion: the calculation of the exact binomial expansion is based on the formula

$$p_y = \left(\frac{1}{2}\right)^N \left(\frac{1}{2}\right)^N \frac{N!}{(N! - y)!y!} \tag{3.1}$$

Where p_y, the probability of making the correct response; y, number of correct responses; and N the total number of responses.

Calculating the appropriate values, even with modern computing power tends to be quite tedious. However, Roessler et al. (1978) published a series of statistical tables based on the binomial expansion and these are widely used in the sensory industry (see Lawless and Heymann, 2010). They also published equations that excellently approximate the binomial expansions for different guessing probabilities, which we provide here since they are simple to use and would be readily available if one has no access to the Roessler tables.

The approximation equation used for the *duo–trio test* is

$$X = \frac{\left(z\sqrt{n} + n + 1\right)}{2} \tag{3.2}$$

Where X, the minimum number of correct judgments if X is a whole number or the next higher integer if X is not a whole number; N, total number of responses; z, z-value for $p = 0.05$ ($z = 1.645$).

Using the equation is quite simple; for example, 50 panelists performed a duo–trio test and 33 correctly matched the reference. Based on this result, was there a significant perceptible difference between the two products?

$$X = \frac{(1.645 \bullet \sqrt{50} + 50 + 1)}{2} = 31.3159$$

One would need at least 32 correct responses to find a significant difference. In the example, there were 33 correct matches and thus we could say that at a probability of 95% or more there was a significant perceptible difference between the two products.

The equation used for the *triangle and tetrad tests* is

$$X = 0.4714 \cdot z \cdot \sqrt{n} + \frac{(2 \cdot n + 3)}{6} \tag{3.3}$$

Where X, the minimum number of correct judgments if X is a whole number or the next higher integer if X is not a whole number; N, total number of responses; z, z-value for $p = 0.05$ ($z = 1.645$).

Again, it is simple to do the calculation; for example, 25 panelists performed a tetrad test and 12 correctly matched the reference. Was there a significant perceptible difference between the two products?

$$X = 0.4714 \cdot (1.645) \cdot \sqrt{25} + \frac{(2 \cdot (25) + 3)}{6} = 12.7106$$

Thus, one would need 13 correct responses to find a significant difference. In the example, there were 12 correct responses and thus we could say that at a probability of 95% or more there was no perceptible difference between the 2 products.

z-tests: the advantage of the z-test over the binomial expansion and the χ^2-test is that one can calculate the exact probability resulting from the number of correct judgments. The equation associated with the z-test is

$$Z = \frac{(X - n \cdot p - 0.5)}{\sqrt{n \cdot p \cdot q}} \tag{3.4}$$

Where X, number of correct responses; N, total number of responses; p, probability of correct decision by chance; for duo–trio test $p = ½$ and for the triangle and tetrad tests $p = ⅓$; $q = (1-p)$; 0.5 is a continuity correction.

In the tetrad example described earlier, where 12 of 25 panelists correctly grouped the samples, the Z-test would come to:

$$Z = \frac{(12 - 25 * 0.33333 - 0.5)}{\sqrt{(25) * 0.33333 * 0.66667}} = 1.34$$

The cumulative probability associated with a z-value of 1.34 is 0.9099 and thus the probability of having this many correct responses is 1 – 0.9099 or 0.0901. Again, we can say that at a probability of 95% or more there was no perceptible difference between the two products.

Thurstonian scaling: we will very briefly describe the use of Thurstonian scaling in the analysis of difference data but will refer the reader to much more indepth discussions on the use of this technique in sensory data analyses summarized by Lawless

(2013a), Ennis et al. (2014), as well as by Brockhoff and Christensen (2010). The Thurstonian model assumes that perceptual noise (due to potential physical variation in the sample, as well as at the receptor level or due to neural noise) changes from moment to moment as the panelist is evaluating the sample. For example, in Fig. 3.1, the panelist is comparing two samples to determine which sample is less intense in a specified attribute. If panelist 1 momentarily perceives intensity a_1 for sample "a" and b for sample "b" she would say that "b" is more intense than "a"—and be correct. However, if panelist 2 momentarily perceives intensity a_2 for sample "a" and b for sample "b" he would say that "b" is less intense than "a"—and be incorrect. It is assumed that the perceptual noise is normally distributed, and that the noise can be both decreased (e.g., by training panelists, see Ishii et al., 2013) or increased due to sensory fatigue and memory demands (Garcia et al., 2012). The difference between the sample means, using Thurstonian theory, is called δ and it is measured in terms of perceptual standard deviations of the sample distributions. From experimental data it is possible to calculate d' as an estimate of δ.

The associated psychometric functions show the relationship between δ and the number of correct responses for each discrimination test method. Thus it is possible to calculate d'. The psychometric functions for most discrimination tests are described in Lawless (2013a), and simplified functions for the triangle and specified tetrad tests are outlined in Bi and O'Mahony (2013). Additionally, Lawless (2013a), as well as Bi and O'Mahony (2013) show extensive tables that are simple to use by a practising sensory scientist.

The advantage of the Thurstonian model is that one can compare the results from different discrimination tests with one another since d' is essentially invariant

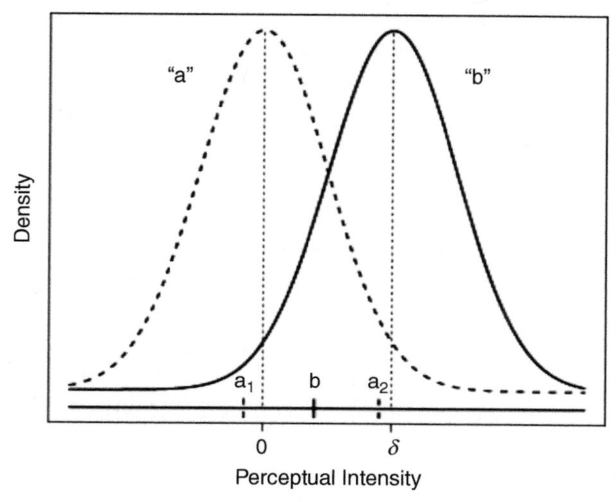

Figure 3.1
Possible percepts for a two sample difference test
a_1 and a_2 are two momentary perceptions of sample "a" and b is a momentary perception of sample "b." *Reprinted with permission from Ennis, J.M., Rousseau, B., Ennis, D.M, 2014. Sensory difference tests as measurement instruments: a review of recent advances. J. Sens. Stud. 29, 89–102*

across methods. As we will see in the next section, this allows us to speak with more confidence about the statistical power associated with each test.

3.1.2 Statistical power of discrimination tests

When designing a sensory experiment the sensory scientist must always keep the statistical power of the study in mind. Without sufficient power, there is no point in actually doing the study. In experimental design, there is always a null hypothesis (H_0) that is stated. In sensory studies, this hypothesis is usually that there is no significant perceptible difference between two samples or among three or more samples. The alternate hypothesis (H_A) is usually that the two samples are perceptibly different from each other or that at least one of the three or more samples differs perceptibly from the others. Traditionally, we choose an alpha (α) probability of 5%. Specifically by doing this we choose to minimize our type I error. Formally, the type I error is the probability of rejecting the null hypothesis (H_0) when it is in fact true. However, there is also a type II error and this is formally the probability of not rejecting (accepting) the null hypothesis when it is actually false. It is not possible to accept a null hypothesis and thus we need to be careful in using this word in this context. The type II error is usually listed as the beta (β) level and it is very rarely explicitly given by sensory scientists. Statistical power is connected to the type II error as it is defined as $1-\beta$. This is not problematic when the experiment leads to an α level that allows one to reject the null hypothesis (i.e., that there are perceivable differences between the samples). In this situation, the study had enough power and a significant perceptible difference was found. However, in situations where the null hypothesis cannot be rejected, the sensory scientist is faced with two potential reasons for the outcome. The first is that the null hypothesis is accurate and there are not perceptible differences between the samples. The second is that there was not sufficient power in the study to reject the null hypothesis even though there actually are significant perceptual differences between the samples. To answer which of these two options is correct the sensory professional had to have designed the study with an explicit β level of probability, which would allow the calculation of the power associated with the study, since statistical power is defined as $1-\beta$.

The statistical power is affected by three factors, namely, the chosen α level, the size of the difference between the samples, and the number of observations in the study. The chosen α-level affects the statistical power because as the α level decreases, the β level increases and thus, the power of the study decreases. This is why the experimenter must choose an α level prior to the beginning of the experiment (usually 5%). The second factor affecting statistical power is the difference between the samples, the smaller the difference between the samples, the lower the signal to noise ratio, which inversely leads to a lower statistical power. In conducting an experiment these two factors cannot be changed, both the α level (most scientific publications require a type I error of 5%) and the size of the difference between the samples are fixed. The only factor affecting the statistical power of the study that the experimenter can control is the number of observations. In effect, the mantra should be that "Power is people," that means as the number of panelists increases the statistical power increases as well. Bausell and Li (2002), as well as Lawless (2013b) discuss power and its importance in similarity and equivalence testing, respectively, in much more depth.

Calculating the power of a study can be done in several ways, for example, there are tables (Ennis, 1993; Schlich, 1993) available that aid the practitioner in calculating the power of a sensory discrimination test both during the design stage (a priori), which is ideally when one should do this, or post hoc when the null hypothesis could not be rejected (a less ideal time to do the calculation). There is also a R-package (sensR, Christensen and Brockhoff, 2015) that allows these calculations and we will show an example in Case Study 1. Please keep in mind that if the samples are perceptibly not very different and one wants/needs a reasonable statistical power (80% or more) then the number of panelists quickly becomes astronomically high!

3.2 Intensity ranking tests

Intensity ranking, unlike hedonic ranking (see Section 3.6.2) was rarely used in sensory science but with the increased popularity of flash profiling (Dairou and Sieffermann, 2002) it is now seen more frequently. In intensity ranking the panelists are asked to rank the samples from lowest to highest intensity of the specified attribute. In most cases, panelists are not allowed to have ties in the rankings, since ties complicate the data analyses. If the sensory practitioner allows ties then they have to be very careful in their choice of data analysis technique. Ranking intensities is a relatively simple and intuitive task for the panelists since most would have been ranking themselves as to height while lining up in elementary school. The task is simple but the panelists should be trained to correctly identify the sensory attribute(s) being ranked. In Section 3.3 we discuss training for descriptive analyses and similar training methods would be applicable in the case of intensity ranking.

In a ranking test, panelists receive all the samples simultaneously, usually coded with three-digit random numbers, and are then asked to rank the samples from highest to lowest (or lowest to highest) intensity without using ties. Because all samples will be evaluated a number of times in a single session to determine their exact rank, the process can be very fatiguing to the panelists. Therefore, especially in the cases of "challenging" (challenging due to high alcohol content and/or high tannin content and/or high sugar content) alcoholic beverages, panelists should only be asked to rank three to at most five samples. Since there is no method to statistically compare panelists ranking across replication, nor to determine a panelist's inherent variation relative to other panelists, replication is not done in ranking tests. Therefore, to gain sufficient statistical power the number of observations must be relatively large, with 50–100 frequently used as a rule of thumb.

The ranking tests are usually analyzed by using the Newell and MacFarlane (1987) tables or by using Friedman's analysis for ranks (Sheskin, 2007). The Newell and MacFarlane tables list the critical value and rank sums differing by this amount or more would be significantly different at the 5% α level (Table 3.1).

Table 3.1 Critical absolute rank sum differences for "all treatments" comparisons at 5% level of significance

Number of observations	Number of samples									
3	6	8	11	13	15	18	20	23	25	28
4	7	10	13	15	18	21	24	27	30	33
5	8	11	14	17	21	24	27	30	34	37
6	9	12	15	19	22	26	30	34	37	42
7	10	13	17	20	24	28	32	36	40	44
8	10	14	18	22	26	30	34	39	43	47
9	10	15	19	23	27	32	36	41	46	50
10	1	15	20	24	29	34	38	43	48	53
11	11	16	21	26	30	35	40	45	51	56
12	12	17	22	27	32	37	42	48	53	58
13	12	18	23	28	33	39	44	50	55	61
14	13	18	24	29	34	40	46	52	57	63
15	13	19	24	30	36	42	47	53	59	66
16	14	19	25	31	37	42	49	55	61	67
17	14	20	26	32	38	44	50	56	63	69
18	15	20	26	32	39	45	51	58	65	71
19	15	21	27	33	40	46	53	60	66	73
20	15	21	28	34	41	47	54	61	68	75
21	16	22	28	35	42	49	56	63	70	77
22	16	22	29	36	43	50	57	64	71	79
23	16	23	30	37	44	51	58	65	73	80
24	17	23	30	37	45	52	59	67	74	82
25	17	24	31	38	46	53	61	68	76	84
26	17	24	32	39	46	54	62	70	77	85
27	18	25	32	40	47	55	63	71	79	87
28	18	25	33	40	48	56	64	72	80	89
29	18	26	33	41	49	57	65	73	82	90
30	19	26	34	42	50	58	66	75	83	92

Table 3.1 Critical absolute rank sum differences for "all treatments" comparisons at 5% level of significance (*cont.*)

Number of observations	Number of samples									
35	20	28	37	45	54	63	72	81	90	99
40	21	30	39	48	57	67	76	86	96	106
45	23	32	41	51	61	71	81	91	102	112
50	24	34	44	54	64	75	85	96	107	118
55	25	34	46	56	67	78	90	101	112	124
60	26	37	48	59	70	82	94	105	117	130
65	27	38	50	61	73	85	97	110	122	135
70	28	40	52	64	76	88	101	114	127	140
75	29	41	53	66	79	91	105	118	131	145
80	30	42	55	68	81	94	108	122	136	150
85	31	44	57	70	84	97	111	125	140	154
90	32	45	58	72	86	100	114	129	144	159
95	33	46	60	74	88	103	118	133	148	163
100	34	47	61	76	91	105	121	136	151	167

Source: Adapted with permission from Newell, G.J., MacFarlane, J.D., 1987. Expanded tables for multiple comparison procedures in the analysis of ranked data. J. Food Sci. 52, 1721–1725.

For example, 25 panelists ranked 3 white wines (X, Y, Z; spiked with various levels of tartaric acid) for perceived sourness with 1 = most sour and 3 = least sour.

Panelist	X	Y	Z
1	1	2	3
2	1	2	3
3	3	1	2
4	2	3	1
5	3	2	1
6	1	3	2
7	1	3	2
8	1	2	3
9	2	1	3

(*contiuned*)

Panelist	X	Y	Z
10	1	2	3
11	1	3	2
12	1	3	2
13	2	3	1
14	1	2	3
15	1	3	2
16	3	2	1
17	1	2	3
18	3	2	1
19	1	2	3
20	2	3	1
21	1	3	2
22	1	3	2
23	3	2	1
24	2	3	1
25	1	2	3
Rank sums	40	59	51

According to the rank sum table the critical value for 25 observation and 3 samples is 17. Thus Sample X was significantly more sour than Sample Y (difference in rank sum of 19) but Samples X and Y were not significantly different from Sample Z (difference in rank sum of 11 and 8, respectively) in perceived sourness. We can summarize this result as follows where rank sums with the same superscript do not differ significantly:

Wine	X	Y	Z
Rank sums	40^{a}	59^{b}	$51^{a,b}$

These data can also be analyzed using Friedman's analysis for ranks. This analysis is based on a χ^2 distribution.

$$\chi^2 = \left(\frac{12}{[K \bullet (J) \bullet (J+1)]} \bullet \sum_{j=1}^{J} T_j^2 \right) - 3 \bullet K \bullet (J+1)\chi^2 = \left(\frac{12}{[K(J)(J+1)]} \left[\sum_{j=1}^{J} T_j^2 \right] \right) - 3K(J+1)$$

Where K, number of panelists; J, number of samples; T_j, rank sums; degrees of freedom for $\chi^2 = (J-1)$.

Using the data described earlier we calculate the χ^2-value

$$\chi^2 = \left(\frac{12}{[K(J)(J+1)]} \left[\sum_{j=1}^{J} T_j^2 \right] \right) - 3K(J+1)$$

If the calculated χ^2-value is significant at the required probability level (usually $\alpha = 5\%$) then one can calculate the least significant ranked difference (LSRD) to determine which samples differ significantly.

$$\text{LSRD} = t\sqrt{\frac{J \cdot K \cdot (J+1)}{6}}$$

Where K, number of panelists; J, number of samples; t, the critical value at the specified α-value (usually 5%); and degrees of freedom, infinity.

$$\chi^2 = \left(\frac{12}{[25(3)(3+1)]} [(40)^2 + (59)^2 + (51)^2] \right) - 3(25)(4) = 7.28$$

Critical χ^2-value for 2 degrees of freedom and $\alpha = 5\%$ is 5.99, thus the rank sums of the samples differ significantly. Now, we can calculate the LSRD value.

$$\text{LSRD} = 1.645\sqrt{\frac{3*25*4}{6}} = 11.63$$

In this example, where the sensory scientist knows that the wines increase in amount of added acid, the t-value is one-tailed and $t = 1.645$ at degrees of freedom = infinity. The LSRD is 11.63 and thus Sample X was significantly more sour than Sample Y but Samples X and Y were not significantly different from Sample Z in perceived sourness. As before we can summarize this result as follows where rank sums with the same supercript do not differ signifcantly.

Wine	X	Y	Z
Rank sums	40^{a}	59^{b}	$51^{a,b}$

In this specific example the Friedman's LSRD and the Newell and MacFarlane tables led to the same conclusions but the LSRD is less conservative than the Newell and MacFarlane tables and the results are not always equivalent.

3.3 Descriptive analyses

The descriptive analysis (DA) methods, also called profiling methods, are the gold standard methods in analytical sensory data collection. In this section we will briefly describe these methods but for interested readers we would also suggest that they see Lawless and Heymann (2010), Meilgaard et al. (2007), Stone et al. (2012), or Heymann et al. (2014).

In DA methods, a group of panelists are trained to use well-defined attributes to describe the products under consideration. The panel can be trained using a consensus method, where the panelists themselves generate the attributes that will be used in the sample evaluation, or by using a ballot method where the attributes are predetermined by the sensory scientist. In both cases it is important that the panel leader provides the panelists with reference standards to anchor and align their concepts of each attribute. These reference standards, preferably actual compounds or products indicative of the specified attribute, will also then act as translation devices to the world at large.

Screening: in Chapter 1 we extensively described the natural variation in the human senses. It is likely that every panelist in a training for a descriptive panel on alcoholic beverages is specifically nondiscriminative to one or more of the attributes under consideration. Conversely, it is also likely that every panelist is more sensitive for one or more attributes than the panel average. It is our belief that screening a panel to ensure that no one is specifically insensitive or overly sensitive to any of the attributes under consideration is a very large (and probably futile) undertaking. After over 35 years of experience training sensory panels we believe that all panelists bring something to the table and add to the data as long as they are motivated and willing participants. Therefore, we very rarely exclude panelists based on screening exercises, in fact we rarely do screening exercises. The exception to this rule would be if the entire project is based on color evaluation say of red wines. In this case we would use the Ishihara plate test (available from reputable optometry suppliers and online) to eliminate panelists who are red color-blind. However, if we were doing a DA of red wines and the panel chose red color as one of their multiple attributes then we would not exclude the color-blind panelists since they would provide information on the aroma, flavor, taste, and mouthfeel attributes.

Training: during the consensus-training period the panelists are exposed to the samples that are under study. They then, through discussion with one another, reach a consensus as to which attributes are present, and through the use of reference standards, they agree on the sensory concept associated with the attribute. During ballot training the panelists are also exposed to the samples that are in the study, but they are given a set of attributes, with reference standards, that will be used in the study. The panelists using the consensus method are creating a new language of attributes to describe the samples, while the panelists using the ballot method are taught a new language of attributes to describe the samples. It is usually easier and faster to use consensus rather than ballot training. However, the key to training a descriptive panel is the interactions among panelists about the sensory attributes associated with the samples. Therefore training should occur in groups, not one on one. A side effect of having the panelists interact with all the potential samples is that it allows them, individually, to realize the range of differences associated with each attribute and this, in a sense sets the edges of the scale for each panelist.

Training sessions usually last approximately 1 h and we usually schedule three to four a week. The length of training is to a large extent panelist and product dependent. Panelists that have participated in previous descriptive analyses will take less time to train than panelists that are totally new to the process of sensory evaluation. Some products are simpler and panelists will determine the sensory attributes needed to describe the products with less training while other products have more subtle differences and are therefore more challenging for the panelists. We have found that length of training typically increases with complexity of the sample with relatively easy products, such as vermouth requiring less time than say a white wine panel.

Sample type	Typical number of 1-h training sessions
Vermouth	4–5 sessions
White wine	5–8 sessions
Gin or tequila (aroma only)	6–7 sessions
Beer	6–8 sessions
Whiskey (aroma and in-mouth)	8–10 sessions
Red wine	7–10 sessions (less for wines that are very different, more for wines that subtly differ)

Reference standards: reference standards allow the panelists to align their concepts of the attributes during training and they also allow outsiders to recreate the reference standards and thus, to get an understanding of the concept used by the panel. There are three types of reference standards and examples of each type are shown in Table 3.2. In certain cases, such as aroma and taste attributes it is possible to use the actual compound in solution as the reference standard. In these cases, the sensory scientist must ensure that the compound is safe for human consumption—and pure. In many cases however, either due to human subjects' safety rules or because the attribute is not due to an impact compound it is not possible to use chemical compounds as reference standards. In these cases the reference standards (usually for aroma terms only) are made using actual products available in the grocery store. A less-ideal solution is to verbally define the attribute, as this solution leaves the possibility that not all panelists understand and use the words in the same context.

Data collection: once the panel is trained the performance should be tested. Usually one has the panelists evaluate a subset of samples in at least duplicate. The subset should be chosen to include both the samples with the smallest potential differences and those with the largest. The data should be analyzed and if all panelists seem to be performing well, one can continue with the actual data collection process. If, on the other hand, the entire panel is not performing well on one or more attributes, then training should continue. The more usual situation is that one or more panelists have problems with different attributes and in this case some one-on-one additional training usually resolves the issues.

Table 3.2 Examples of reference standards used in descriptive analyses of alcoholic beverages

Modality	Attribute	Reference standard	Reference type
Taste	Bitter	1 g/L caffeine	Actual compound
Taste	Umami	16 g/L monosodium glutamate	Actual compound
Taste	Sour	2.5 g/L tartaric acid	Actual compound
Taste	Sweet	10 g/L sucrose	Actual compound
Taste	Salty	3 g coarse Kosher salt in 1000 mL water	Actual compound
Aroma[a]	Sweet sweaty passionfruit	2000 ng/L 3-mercaptohexyl acetate	Actual compound
Aroma[a]	Bell pepper (capsicum)	1000 ng/L 2-methoxy-3-isobutylpyrazine	Actual compound
Aroma[a]	Cat urine/boxwood	1000 ng/L 4-mercaptomethylpentane	Actual compound
Aroma[a]	Passion fruit skin/stalk	2000 ng/L 3-mercaptohexan-1-ol	Actual compound
Mouthfeel	Viscous	2 g/L carboxymethyl cellulose	Actual compound
Mouthfeel	Astringent	840 mg/L alum	Actual compound
Aroma	Artificial fruit	1/8 tsp KoolAid Red Cherry in 150 mL base red wine[b]	Product
Aroma	Dark fruit	2 frozen blackberries, thawed and crushed+1 tsp black currant jam	Product
Aroma	Dried fruit	2 dried figs, 3 prunes, 20 raisins in 60 mL base wine	Product
Aroma	Cardboard	Five 1 $in.^2$ pieces of corrugated cardboard in 60 mL water	Product
Aroma	Earthy/moldy	500 mg of wood+800 mg of freshly cut portobello mushroom lamellae	Product
Aroma	Sulfur	1/2 hardboiled egg yolk stored at room temperature for at least 24 h	Product
Aroma	Citrus/cola	2 mL Shasta Cola+10 mL 5% ethanol solution	Product
Mouthfeel	Cooling	15 drops eucalyptus extract+500 mL water (instruct panelists to pinch nose to eliminate aroma/flavor)	Product
Mouthfeel	Hot	250 mL vodka+750 mL water	Product
Mouthfeel	Burning	Defined as the physical sensation of tingling or numbing of the tongue	Verbal definition
Mouthfeel	Creamy mouthfeel	Defined as a smooth, velvety, creamy feeling in the mouth	Verbal definition

[a]*From Lund, C.M., Thompson, M.K., Benkwitz, F., Wohler, M.W., Triggs, C.M., Gardner, R., Heymann, H., Nicolau, L., 2009. New Zealand Sauvignon blanc distinct flavor characteristics: sensory, chemical, and consumer aspects. Am. J. Enol. Vitic. 60, 1–1.*

[b]*The base wine is a neutral bag-in-the-box wine of the appropriate color (red in the case of a red wine study and white in the case of a white wine study).*

Data collection usually occurs in individual booths under controlled environmental conditions (temperature, humidity, lighting, etc.). Since the advent of inexpensive computers it is more and more likely that the data will be collected using a electronic data collection system. However, many companies that do only a few DA studies a year still use paper ballots. Based on some research by Swaney-Stueve and Heymann (2002) paper and computerized ballots can be used interchangeably for DA data collection. As an aside, Savidan and Morris (2015) recently showed that this is also true for napping or projective mapping (PM) data (discussed later in this chapter).

Data analysis: it is not the intent in this chapter to exhaustively describe how DA data should be analyzed. Our intent is to provide the reader with some basic information about the topic and then to identify a number of publications that would be of use for further reading. Case studies 2, and 3 also show how to analyze DA data using R. DA data are usually first analyzed by univariate analysis of variance (ANOVA) to determine which of the rated attributes discriminate among the samples. Usually, once this information is determined, a mean separations technique is used to show which samples differed significantly from one another. The next step in the data analysis is then to do a multivariate analysis of the data to visually show the differences among the samples in a two- or three-dimensional graph. There are a number of appropriate multivariate techniques that could be used but we will only briefly discuss two—canonical variate analysis (CVA) and principal component analysis (PCA). If the descriptive data has to be evaluated in conjunction with chemical or other data then there are once again a number of appropriate techniques but we will briefly only describe one—multiple factor analysis (MFA).

ANOVA: in an ANOVA the variance associated with the data is distributed across the sources of variation in the data set. In the simplest DA data sets the main sources of variation, the so-called main effects, would be the samples, the panelists, and the replications. Additional sources of variation would be the interaction effects between the samples and panelists, samples, and replications, as well as the interaction between panelists and replications. Please see Lea et al. (1997) and Næs et al. (2010) for more in depth discussions of ANOVA, as well as Case Study 2 for the R-code associated with an ANOVA. If the ANOVA indicates that there is a significant difference among the samples, then the sensory scientist would perform a means separations technique to determine which of the samples differed from one another. Traditionally, Fisher's protected least significant difference (LSD) test is used but other more conservative mean separations techniques, such as Tukey's test, the Bonferroni test, Duncan's multiple range, and so on also exist (Gacula and Singh, 1984). The LSD can only be calculated once the ANOVA showed that the samples differed significantly. The LSD can be calculated using the following equation:

$$\mathrm{LSD} = t\sqrt{\frac{2MSE}{N}}$$

Where t, the two-tailed value for the specified α level, usually 5% with the degrees of freedom associated with the error term; MSE, mean square error value; N, number of observations used to calculate the mean values for the samples. This equation assumes a balanced data set where all panelists evaluated all samples and equal number of times, without any missing data points.

CVA: a CVA uses the raw sensory DA data to create linear combinations of the data set variables to maximize the variance ratio (variance between samples to variance within samples) explained and the usual output is a two- or three-dimensional graph of the data space (Heymann and Noble, 1989; Peltier et al., 2015). The initial step is a one-way (treatment only) multivariate analysis of variance (MANOVA) of the data set (Heymann et al., 2014). If the MANOVA is significant at the desired α level, the CVA can then be calculated. In a sense the CVA is a multidimensional "mean separation" technique for the MANOVA, in analogy to the LSD being a mean separation technique for the ANOVA. Since the CVA is performed on raw data it is possible to calculate the 95% confidence ellipses around each sample in the two- or three-dimensional output. Fig. 3.2 shows an example of a CVA on the wines associated with Table 3.3 (extracted from Hopfer and Heymann, 2014).

PCA: traditionally the PCA is performed on the mean data (treatment by variable matrix). The PCA creates linear combinations of the variables in the data set to maximize the variance explained and again the usual output is a two- or three-dimensional graph of the data space (Peltier et al., 2015). In general, if one has access to raw data then the CVA performs best but as shown by

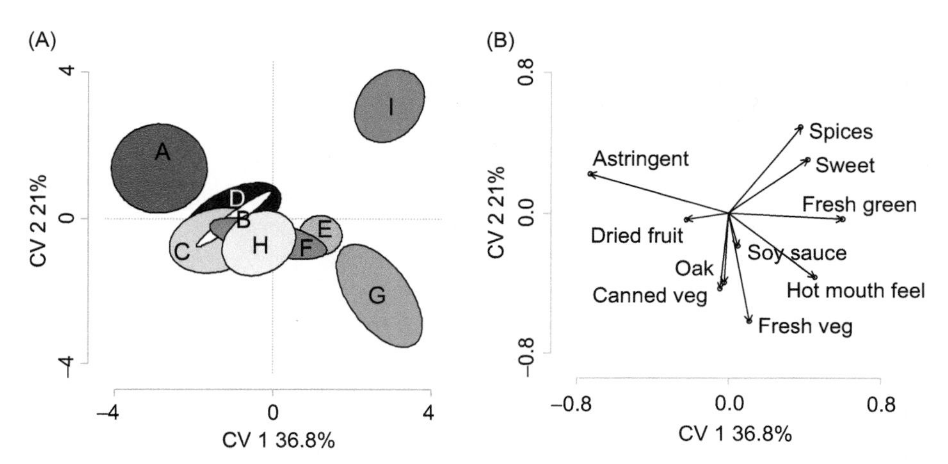

Figure 3.2
(A) Canonical variates *(CV)* analysis of the 10 significant sensory attributes for Cabernet Sauvignon wines across the nine California wine regions. CV scores and their 95% confidence intervals for CV 1 and CV 2. See Table 3.3 for information on regions and (B) CV loadings for CV 1 and CV 2 for the 10 significant sensory attributes for Cabernet Sauvignon wines across the nine California wine regions.

Table 3.3 California wine regions shown in Fig. 3.2A

Region	Location[a]	US$[b]
A	North Coast	25.33
B	Sonoma	34.66
C	Napa	53.67
D	Greater Bay Area	30.33
E	North Central Coast	12.00
F	South Central Coast	17.50
G	South Coast	43.65
H	Sierra Foothills	28.50
I	Lodi	23.00

[a]*Locations are within the State of California, USA.*
[b]*Average retail price per 750 mL bottle.*

Peltier et al. (2015) PCA is a reasonable alternative in most instances, however, we prefer to use raw data to create the multidimensional data spaces for DA data and it is only in the past decade that an alternative to CVA that allows this became reality. Lê and Husson (2008) in their R-package, SensomineR, created a PCA analysis that allows the use of raw data (not averaged) and the calculation of 95% confidence ellipses intervals (see Case Study 3).

When doing a PCA the sensory professional must decide whether to use a correlation or a covariance matrix as input. A correlation matrix must be used when the data set has variables measured on different scales, for example, for a red wine pH may range from 3 to 4.2, alcohol from 10 to 17% (v/V), residual sugar from 0 to 0.5%, tannin ranging from 250 to 800 mg/L. In this case if a correlation matrix is not used the tannin data would "swamp" the other data and the resultant PCA would be nonsensical. On the other hand, DA data usually has the same scale for all attributes and thus a covariance matrix is appropriate as the input matrix.

3.4 Rapid techniques

DA with trained panels remains the gold standard for analytical sensory evaluation but starting in the 1980s sensory scientists have been creating rapid techniques to either replace DA or to use in conjunction with DA. The impetus for these techniques is to eliminate or minimize the time to train the panelists and thus to speed up the data collection. The first of these techniques was free choice profiling (FCP) invented in the 1984 by Williams and Langron (1984). There are now numerous rapid techniques, for example, free sorting (FS, Lawless and Glatter, 1990; Gilbert and Heymann, 1995), flash profiling (Dairou and Sieffermann, 2002), projective mapping (PM, Risvik et al., 1994), Napping (Pagès, 2005), pivot profile (Thullier et al., 2015), polarized sensory positioning (Teillet et al., 2010), check-all-that-apply (Campo et al., 2010), and so forth. In this section we will briefly describe three of these techniques, FCP, FS, and PM. Anyone interested in more extensive descriptions of these and other rapid techniques should consult Delarue et al. (2015), as well as Varela and Ares (2014).

3.4.1 Free choice profiling

Williams and Langron (1984) suggested that it should be possible to have each panelist use his/her own idiosyncratic language to describe the difference among a set of products. Each panelist was shown the entire range of products and was asked to create a list of attributes to describe the differences among the products. This list was then used to make a unique score sheet for each panelist. The panelists were then served the samples monadically and asked to rate each product on each attribute of their unique score sheet. Some sensory scientists using this method only asked for a single replication of each product from each panelist (Elmore and Heymann, 1999; Perrin et al., 2007) whereas others asked their panelists to perform at least two replications (Piggott and Watson, 1992; Cristovam et al., 2000; Tang and Heymann, 2002). The collected data are analyzed by either generalized Procrustes analysis (GPA, Arnold et al., 2007; Donaldson et al., 2012) or by MFA (Pagès and Husson, 2005; Perrin et al., 2007).

As discussed by Piggott and Watson (1992) some of the disadvantages of FCP are (1) panelists can struggle with generating descriptors that differentiate among the samples, (2) two or more panelists that use the same word to describe different sensation, or (3) panelists that use different words to describe the same sensation. Perrin et al. (2007) showed that wine experts used the word "vegetal" to describe widely divergent odors, such as bell peppers, grass, and asparagus. These authors compared FCP to DA for French Loire Valley wines and found that the two methods were in agreement for the first axis of the space but less so for the second. They concluded that DA is more useful when the differences among samples are subtle and when the use of defined reference standards would anchor the attribute percepts. Elmore and Heymann (1999) comparing FCP and DA found that the perceptual maps for the two methods were quite similar and that the FCP panelists used more explicit attributes to describe the differences among the soda brands. It should be noted that in this study the samples were photographs of soda cans and that the brands represented were very familiar to the American panelists performing the study—there were not many subtle differences! The FCP was extensively used but seemed to fall out of favor in the early 2000s. In 2002 Dairou and Sieffermann (2002) published a variant of the FCP, which they named flash profile (FP). The difference between FCP and FP is that the panelists do not rate the perceived intensity of each attribute for each sample. Instead they rank the samples from lowest to highest intensity.

In the literature FP, panelists perform between one and five replicate evaluations (Tarea et al., 2007; Valentin et al., 2012; Dehlholm et al., 2012; Dairou and Sieffermann, 2002). FP data are typically analyzed by ANOVA, even though these data are clearly nonparametric in nature. Some authors used GPA (Tarea et al., 2007) or MFA (Blancher et al., 2007) to analyze the FP data. The FP method was created for experts or panelists that are very familiar with the product and its descriptors (Valentin et al., 2012) and thus is not recommended for use with untrained panelists. Additionally, since the panelists have to rank all the products on each attribute the method will not work well with very fatiguing products.

3.4.2 Free sorting

Sorting (or free sorting as it is sometimes called) has been used extensively in sensory data collection since Lawless and Glatter (1990) first used it for odor sorting. The process is very simple. The panelists receive all the samples simultaneously and are asked to sort them into similarity groups. They are usually instructed to make at least two groups and no more than the total number of samples minus one groups, so for example, if they are asked to sort ten samples they can make between 2 and 9 groups. Panelists use their own criteria to determine how they would sort the samples. Usually panelists will perform only one replication of the sorting task but there are published examples where more replications were performed (Mielby et al., 2014; Gilbert and Heymann, 1995). The data are compiled by determining the number of times the same samples were sorted together and this forms a similarity matrix, either an overall matrix across all panelists or a by panelist version (Fig. 3.3). Traditionally, the overall similarity matrix

Panelist 1

P2, P4, P5, P7

P1, P6

P3

Panelist 2

P3, P1, P6

P2, P4, P5, P7

	P1	P2	P3	P4	P5	P6	P7
P1	1	0	0	0	0	1	0
P2	0	1	0	1	1	0	1
P3	0	0	1	0	0	0	0
P4	0	1	0	1	1	0	1
P5	0	1	0	1	1	0	1
P6	1	0	0	0	0	1	0
P7	0	1	0	1	1	0	1

	P1	P2	P3	P4	P5	P6	P7
P1	1	0	1	0	0	1	0
P2	0	1	0	1	1	0	1
P3	1	0	1	0	0	1	0
P4	0	1	0	1	1	0	1
P5	0	1	0	1	1	0	1
P6	1	0	1	0	0	1	0
P7	0	1	0	1	1	0	1

Figure 3.3
Examples of the similarity matrices obtained in a sorting
Two panelists were asked to sort seven products into at least two and no more than six groups. The number of times a sample was sorted into the same group was then coded into a similarity matrix. Please note that the diagonal values are all 1 since a product is always sorted with itself. These individual panelist similarity matrices are the combined to form an aggregate matrix which would be analyzed by MDS or DISTATIS.

from a sorting study was analyzed by multidimensional scaling (MDS, Popper and Heymann, 1996). The MDS creates a two or three-dimensional plot that allows one to see underlying structure in the data set and to determine similarities and differences between products. MDS requires an overall similarity matrix and this masks any panelist-to-panelist variability (Abdi et al., 2007). More recently, DISTATIS analysis, that allows one to determine both individual panelist effects and aggregate effects, have been used (Abdi et al., 2007; Lahne et al., 2016).

Panelists sort all products at the same time, thus, this technique does not work very well with a large number of very fatiguing products or with products, such as sparkling wines that will change during the time frame of the sorting exercise.

3.4.3 Projective mapping

Projective mapping was first suggested as a rapid sensory technique by Risvik et al. (1994, 1997) but is was not until Pagès (2005) created Napping that it became very popular in the sensory science community. In both these techniques the panelists receive all the samples simultaneously and are asked to place the samples on a sheet of paper (often a 60×40 cm^2 sheet) in such a way that more similar samples are closer together and less-similar samples are further apart (Fig. 3.4). As in sorting, the panelists use their own criteria as to how they

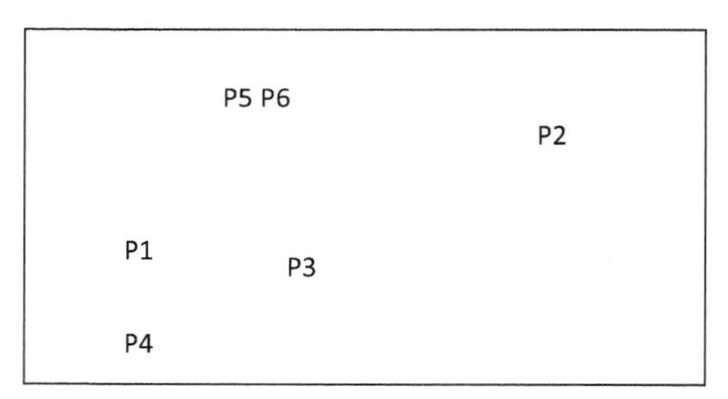

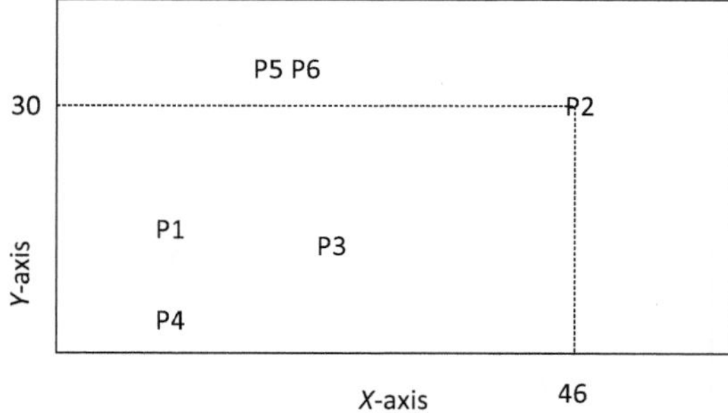

Figure 3.4
In this example of projective mapping the panelist placed six products on the sheet of paper in such a way that products that are similar are close together and ones that a very different a further apart Once all samples had been placed on the paper the position of each product was recorded by measuring the *X*- and *Y*-axis position of each product.

would group and place the samples. Interestingly, in the earliest publications on projective mapping the panelists performed the mapping task in triplicate (Risvik et al., 1997; Barcenas et al., 2004) but in later publications it seems as if only Heymann and coworkers usually have at least duplicate mappings (see e.g., Mielby et al., 2014; Heymann et al., 2014). Once all the samples have been placed on the panelist's paper, the panelist may be asked to describe the product or product groups by writing a few words next to each on the sheet of paper. If this variant of projective mapping is used it is called ultra flash profiling (Perrin and Pagès, 2009).

For data analysis, the X and Y coordinates of each product on each panelist's product map are recorded and used in the multivariate data analysis (Fig. 3.4). Recently, the data analysis method of choice has been MFA but as it is also possible to use GPA (Tomic, 2013).

MFA: MFA or multiple factor was popularized by Pagès and coworkers (Escofier and Pagès, 1990; see e.g., Pagès and Husson, 2005; Lê et al., 2008) as a multivariate technique that allows the comparison of multitable and multiblock data sets. The technique is very useful in the analysis of projective mapping (napping) data (see Case Study 8) but it is also very useful in the simultaneous analysis of sensory and chemical or physical data (see case study 10). The technique, as it is usually applied in sensory analyses can be used when different (or the same) attributes were measured on the same samples. In the case of projective mapping data all panelists provide *x*- and *y*-axis data for each sample on the table cloth—in this case the same attributes were measured for the same samples by multiple panelists. When comparing chemical and sensory data the same samples were measured by DA and say also by headspace-gas chromatography-mass spectrometry. It not usual in sensory data but MFA can also be used when the same attributes are measured on different samples (Abdi et al., 2013). The MFA can be viewed as a PCA of PCAs in that the technique calculates individual PCAs for each data set and then the data set is "normalized" by dividing each element of each data set by the first singular value of the PCA for that data set. Next the normalized data sets are combined into a single data set and this combined data set is analyzed by a PCA. The output is factor scores for the samples and loadings for each of the variables in the combined data set. Additionally, the MFA also provides partial factor score for each data set providing the specific viewpoint of each data set.

3.4.4 Comparison of rapid techniques and descriptive analysis

There have been numerous comparisons of the rapid techniques with one another and with classical DA (see e.g., Delarue et al., 2015; Varela and Ares, 2014). In many cases, if the samples have large sensory differences, the techniques work equally well. However, if the intent is to discover more subtle differences then the rapid techniques tend to compare poorly to traditional DA. As can be seen

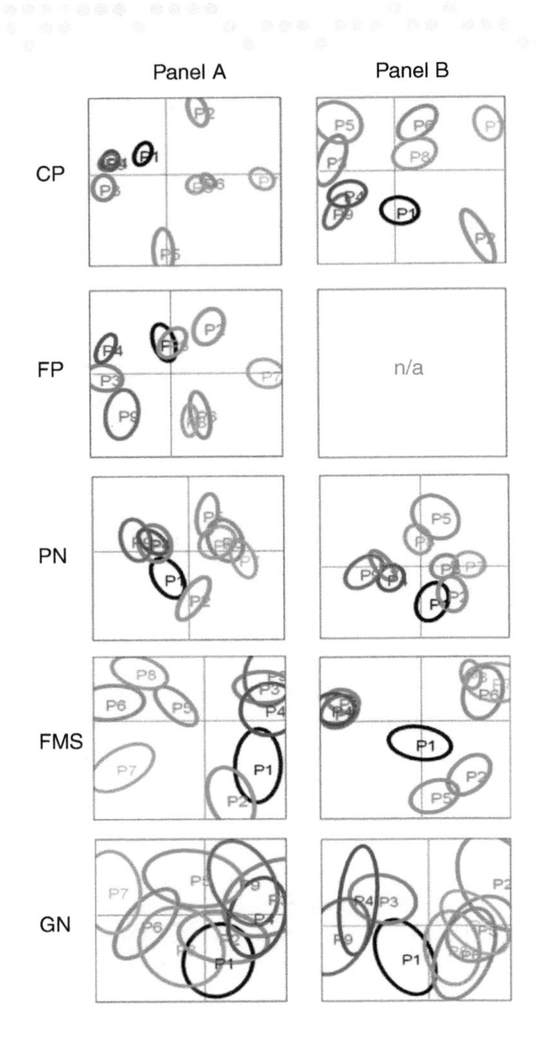

Figure 3.5 Individual configurations for a selection of rapid methods showing each sample's 95% confidence interval
The higher the plot, the more time is spent on sample evaluation. *CP*, Conventional descriptive profile; *FMS*, free multiple sorting; *FP*, flash profile; *GN*, global napping; *PN*, partial napping means. *Reprinted with permission from Dehlholm, C., Brockhoff, P.B., Meinert, L., Aaslyng, M., Bredie, W.L.P., 2012. Rapid descriptive methods—comparison of free multiple sorting, partial napping, napping, flash profiling and conventional profiling. Food Qual. Pref. 26, 267–277*

in Fig. 3.5 the reason for this is that the uncertainty associated with the rapid methods tends to be much larger than for DA.

3.5 Time-dependent techniques

None of the techniques described earlier take into account that the sensory attributes of a product may change as the panelist evaluates the product. A simple example would be ice cream melting in the mouth as the panelists are evaluating the mouthfeel and taste properties of the ice cream. In a wine context, the perceived astringency of a red wine may increase as the panelist moves the liquid around in his/her mouth. In general, in DA when panelists are asked to assess the astringency of a sample they essentially subconsciously "average" the perceived astringency over the time that the wine was in their mouth and give a single rating. However, this "averaging" may hide some very interesting information, and this is where the time-dependent sensory techniques are very helpful. It is entirely possible to obtain discrete time point information from a DA panel. For this, the panel evaluates the perceived astringency of the wine at different time points, for example, immediately before expectoration, then 30 s after and finally 2 min after. However, sometimes the sensory scientist is interested in more continuous time point information and the two techniques that provide this type of information are time–intensity (TI) and temporal dominance of sensations (TDS). We will discuss these in the next two sections.

3.5.1 Time–intensity

Similar to a classical DA, when using the TI approach, panelists are trained to evaluate the specific attributes of interest using reference standards and consensus training. Usually, for alcoholic beverages, the sensory scientist is often more interested in the time course of the taste and mouthfeel attributes and less in the aroma and flavor by mouth attributes. Most TI studies on alcoholic beverages focus on taste and mouthfeel attributes, such as astringency, bitterness, hotness, sourness, and sweetness (see e.g., Pickering et al., 1998; Guinard et al., 1996; Leach and Noble, 1986) rather than aroma and flavor by mouth attributes. In addition, most TI studies ask the panelist to evaluate a single attribute over time—usually using a mouse or joystick to indicate increases and decreases in the perceived intensity of the attribute. There are a few studies where the panelists were asked to evaluate the perceived intensities of two attributes over time, this technique is called dual attribute time–intensity and is very taxing on the panelist and is not frequently used (Duizer et al., 1996, 1997; Kobue-Lekalake et al., 2011).

Fig. 3.6 shows an example of a completed TI curve and Table 3.4 shows all the parameters that can be calculated from this curve. Sensory scientists very rarely, if ever, use all of these parameters. These parameters are frequently called the scaffolding of the TI-curves. Once the sensory scientist has extracted the scaffolding parameters he/she can then analyze the data using ANOVA. There are some

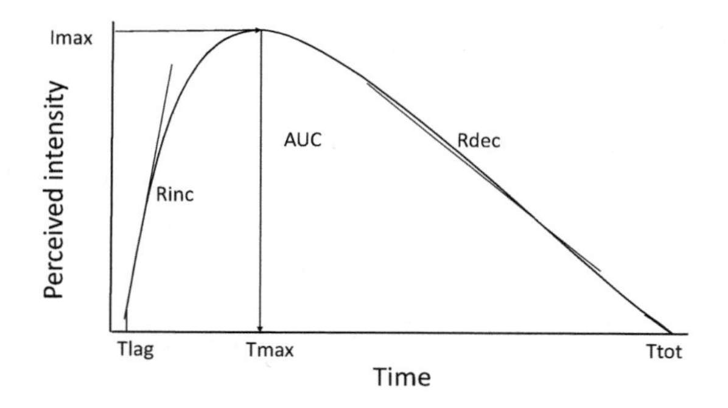

Figure 3.6
A simple representation of a time–intensity curve
See Table 3.4 for the parameters that can be extracted from this curve.

Table 3.4 Some parameters that can be extracted from the TI curve shown in Fig. 3.6

Code	Parameter
I_{MAX}	Maximum intensity value
T_{MAX}	Time to maximum intensity
T_{LAG}	Lag time (time to first nonzero intensity value)
$T_{MAX-LAG}$	Time to maximum intensity adjusted for lag time
T_{LMAX}	Time of last maximum intensity (used to calculate T_{PLAT} or plateau time)
T_{PLAT}	Duration of maximum intensity (time between T_{MAX} and T_{LMAX})
T_{TOT}	Total time
$T_{TOT-LAG}$	Time of sensation from the end of lag time to total time
R_{INC}	Rate of increase (slope from T_{LAG} to T_{MAX})
R_{DEC}	Rate of decrease (slope from T_{LMAX} to T_{TOT})
AUC	Area under the curve
A_{PREMAX}	Area before maximum intensity
$A_{POSTMAX}$	Area after maximum intensity

Adapted from Pickering, G.J., Heatherbell, D.A., Vanhanen, L.P., Barnes, M.F. (1998). The effect of ethanol concentration on the temporal perception of viscosity and density in white wine. Am. J. Enol. Vitic. 49, 306–318.

serious issues with averaging TI-curves and as eloquently explained by Lawless (2013c) none of the methods suggested to mitigate the averaging issues are completely satisfactory. Due to these issues, as well as the potential dumping bias that can occur when panelists are asked to only rate one attribute at a time, as well as the length of time it takes to complete a TI study, the method is not used extensively. Dumping occurs when a sample differs in multiple attributes but the panelists are only asked to rate one attribute. They then tend to subconsciously inflate or deflate the perceived intensity of the rated attribute to compensate for the other attributes differing in the sample (Clark and Lawless, 1994; Sokolowsky and Fischer, 2012).

3.5.2 Temporal dominance of sensations

Unlike the TI method, the TDS technique allows the panelist to indicate which of the specified attributes is dominant at any time during the evaluation. The panelists are provided with a list of attributes (determined through preliminary testing) and are asked to indicate which attributes are sequentially dominant over the course of the evaluation. Dominance is defined as "the sensation that capture's one attention, or the most striking perception" (Labbe et al., 2009) but it must be remembered that dominance is not equivalent to intensity. This could be problematic when trained DA panelists are used to do TDS (Meyners, 2010). It has been suggested that the number of attributes used should be kept low (8–10 according to Pineau et al., 2012) but 6 is ideal since Goupil de Bouille et al. (2010) showed that judges change their rating behavior when there were more than 6 attributes to choose from. Panelists should be trained, using reference standards, to understand the concept of each attribute, and to reach consensus on each attribute (Di Monaco et al., 2014).

Although a relatively new technique, TDS has been used on a number of alcoholic beverages with Sokolowsky and Fischer (2012) studying white wines, Meillon et al. (2009) evaluating partially dealcoholized red wines, Déléris et al. (2011) studying vodka and Vázquez-Araújo et al. (2013) studying beer. TDS should not be viewed as a replacement for DA but rather as a complement (Di Monaco et al., 2014).

The data analyses associated with TDS range from the simple to the more complex but in general there is not yet a true consensus on how to best extract all the potential information from the data. The simplest analysis method is to calculate a dominance rate (percentage of evaluations for which each attribute was selected as dominant) at fixed intervals. Bruzzone et al. (2013) suggested a fixed interval of ½ second. The dominance rate is then smoothed using a polynomial spline function and plotted as a dominance rate–time curve (Fig. 3.7). The

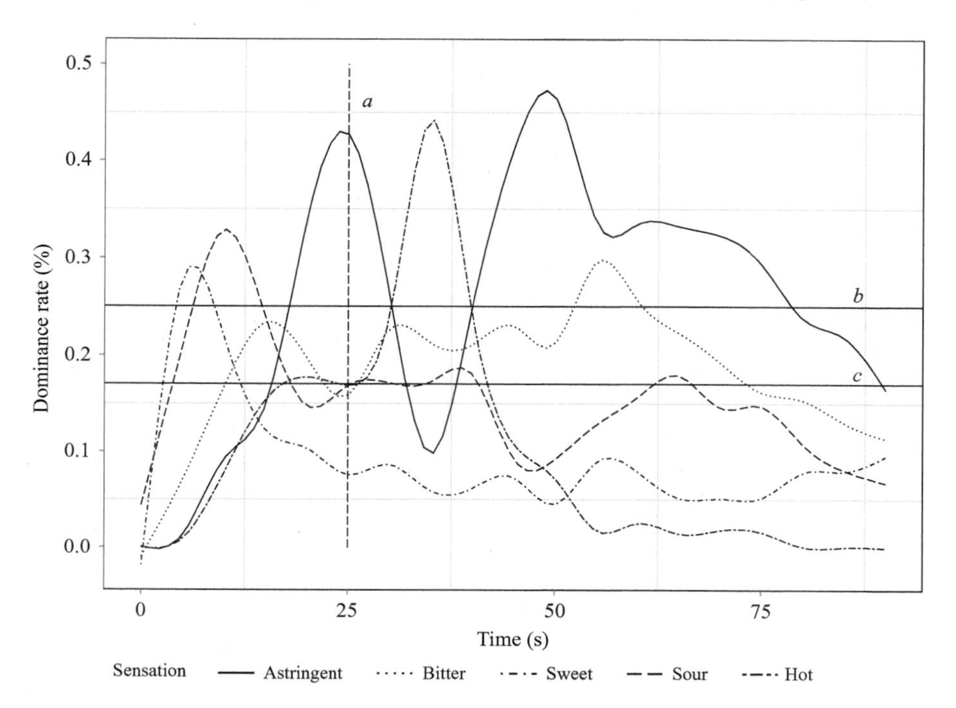

Figure 3.7 An example of a temporal dominance of sensations plot with *a*, expectorate; *b*, significance line; and *c*, chance line

chance level (P_0) and significance level (P_S) is also plotted on the curves. Chance dominance rate is the inverse of the total number of attributes plus one (Labbe et al., 2009), assuming that panelists were allowed to also choose none of the attributes as dominant. The significant dominant rate is the minimum dominance rate of an attribute that is significantly higher than chance. This calculation is based on a binomial test, taking into account the number of observations (panelists and replicates) in the data set (Pineau et al., 2009). Specifically:

$$P_S = P_0 + 1.645\sqrt{\frac{P_0(1-P_0)}{n}}$$

Where P_S, lowest significant dominance rate at any point in time for a TDS curve ($\alpha = 0.05$); P_0, chance dominance rate based on the inverse of the total number of attributes plus one; and n, number of observations (panelists $\times$ replications).

This calculation is based on the confidence interval of a binomial proportion and thus the number of observations (n) should be large enough for $nP_0(1-P_0) > 5$.

More complex data analyses, including the use of ANOVA (Dinella et al., 2013) and PCA (Bruzzone et al. 2013), but the field is still in flux as to the usefulness and validity of these techniques (Di Monaco et al., 2014).

3.6 Consumer sensory analyses

Unlike the analytical sensory methods described in the previous sections, in consumer sensory methods sensory scientists are interested in the subjective responses of the consumers. Here, the data obtained is entirely subjective and because of the inherent noise, many consumers should be used in consumer panels. The key requirement for consumer panelists is that they should be the target consumers of the products under consideration; there is no point in asking hedonic (liking) information on red wine from someone who never consumes red wines. There are a number of different consumer sensory tests: paired preference techniques, hedonic ranking, hedonic scaling, and conjoint analysis. We will briefly describe each of these in the following sections.

3.6.1 Paired preference testing

Prior to doing the paired preference evaluation the sensory scientist has to choose the two products used in the test. These may be an original product compared to a new reformulated product or a comparison of a product against a competitor, and so on. In all cases, the sensory scientist should have previously shown that the products are perceptibly different from one another—for example, by discrimination testing or DA. The reason for this is that asking consumers to indicate their preferences for one product over another when the products are not perceptibly different is a waste of time. Additionally, if the result of a preference test is that there is no significant preference then that does not mean that the samples are not different from one another. Consumers may not significantly prefer eating an apple over a banana but a banana is still not an apple.

In paired preference testing the consumer is asked to indicate which of two products is more preferred. The sensory scientist should ensure that the presentation sequence of the two samples is counter balanced across panelists to prevent a first-sample bias. The paired preference data would be analyzed using the same data analyses as discrimination tests discussed in Section 3.1.1. The only, and very important difference is that now one would be using a two-tailed test since there is no "correct" answer—the consumer could prefer either sample. Therefore the data analysis should account for this by using a two-tailed test.

In most situations, to make the statistical analyses relatively simple, the consumers are required to make a choice and cannot use a *no preference* option. Some researchers feel that there is valid information to be found in a *no preference* option, especially when a large number of the targeted consumers make this choice. However, using the *no preference* option really means that one cannot use the binomial statistical tests that are usually used for discrimination testing and that one should be using a multinomial test. This can lead to the need for external statistical analysis help from a statistician. In an effort to circumvent the need for multinomial data many researchers feel that if one uses the *no preference* option and if relatively few consumers use this option then one can assign these consumers to the two product options in the same ratio as the other consumers (Odesky, 1967). Alternately one could assign these consumers equally between the two preference options (Lawless and Heymann, 2010) or if there are very few *no preference* options then one could ignore them and not use that data (Lawless and Heymann, 2010). However, we think that work by Angulo and O'Mahony (2005) and Marchisano et al. (2003) show clearly that one should not use a *no preference* option.

Usually, preference tests are not replicated on the assumption that consumers are stable in their preferences. However, as shown by Chapman et al. (2006) and Köster et al. (2003) about 50% of consumers may not have stable preferences and thus asking consumers to replicate their evaluations (on a second set of samples) may provide very useful information. Meyners (2007) shows that a χ^2-test gives an easy and powerful way to analyze these replicated data.

3.6.2 Preference ranking

When one wants to determine consumers' preference for more than two products, and probably for fewer than five or six products, then one can perform a preference ranking. The consumers are asked to rank the samples from their least to most preferred, with no ties allowed. Usually consumers find this process fairly simple as long as there are not too many samples and as long as the samples are not extremely fatiguing. These data would then be analyzed using the Newell and MacFarlane (1987) tables and Friedman's rank ANOVA – both of these are described in Section 3.2.

Usually, the sensory scientist would opt to do a hedonic scaling (i.e., consumers are asked to rate the products on a numerical scale) rather than a hedonic ranking (i.e., consumers rank samples from least to most preferred), since the parametric data obtained from a scaling test is more powerful and allows one to analyze the data using many more statistical techniques. Kozak and Cliff (2013) compared hedonic ranking and hedonic scaling and recommended scaling over ranking for these reasons. In our minds, the exception to this would be if the samples are uniformly disliked and thus in a scaling all would be rated the same way but the sensory scientist still wants to find the least disliked sample. In this case doing a forced ranking for preference may identify such a sample where a hedonic scaling would not. Similarly, if the samples are uniformly liked extremely well but the brief is to find the "best"-liked sample then a forced preference ranking may identify this sample.

3.6.3 Hedonic scaling

The nine-point hedonic scale (Fig. 3.8) was developed at the Quartermaster Food and Container Institute of the US Armed Forces, expressly to measure the food preferences of US soldiers (Peryam and Girardot, 1952; Jones et al., 1955; Peryam and Pilgrim, 1957). The nine-point hedonic scale is a bipolar scale with like and dislike on each end and a neutral point in the middle. The scale quickly spread as the preferred hedonic scaling instrument because it works well in showing differences in degree of liking among products and is still likely the most used scale for this purpose This happened despite acknowledged weaknesses, such as lack of linearity of the adjectives (Peryam and Pilgrim, 1957), the likely use of the scale as ordinal rather than interval by the consumers (Moskowitz and Sidel, 1971), the lack of equivalent use of the adjectives among the consumers, and so on. A number of researchers have proposed alternatives to the nine-point hedonic scale, for example, the labeled affective magnitude scale (Schutz and

Please indicate your degree of liking for the product

by encircling the appropriate phrase.

Like extremely

Like very much

Like moderately

Like slightly

Neither like nor dislike

Dislike slightly

Dislike moderately

Dislike very much

Dislike extremely

Figure 3.8 A score sheet using the nine-point hedonic scale

Cardello, 2001), the best/worst method (Marley and Louviere, 2005), the semantically labeled hedonic scale (Lim et al., 2009), the hedonic general labeled magnitude scale (Bartoshuk et al., 2012), and so on. However, in nearly all cases the results of comparison studies showed that the nine-point hedonic scale performed equally well (Kalva et al., 2014; Lim, 2011; Lawless et al., 2009). The one exception to this seems to be best/worst scaling where the nine-point hedonic scale clearly performed better than the best/worst scaling (Mueller et al., 2009; Mielby et al., 2012). For these reasons we suggest that sensory practitioners use the nine-point hedonic scale, since Lim (2011) stated "... when the primary concern of a study is measuring hedonic *differences* among foods, beverages, and consumer products and predicting their *acceptance*, the nine-point hedonic scale has proven itself to be a simple and effective measuring device."

The context of the study can have great effect on the consumers liking of the specific products and the sensory practitioner must take care not to artificially change the context or to bias the context in anyway. As shown by Hein et al. (2012) something as simple as an evoked consumption context (e.g., telling the consumers to imagine drinking a beverage while watching a movie) was enough to lower hedonic scores for a somewhat unfamiliar juice. For additional information about context effects and hedonic scaling please see Lawless and Heymann (2010) or Lim (2011).

Researchers have recommended that the number of targeted consumers needed for a hedonic scaling test range from 50 to at least 100 (Gacula and Rutenbeck, 2006) with Meilgaard et al. (2007) suggesting up to 300 consumers were needed. To calculate the number of required consumers one needs to know the desired α level (usually 0.05), the desired power of the study (say 90%, then the β level would be 0.10), the desired minimum difference (say 0.8) (Bovell-Benjamin et al., 1999 as well as Hough and coworkers used this value for a nine-point hedonic scale. In specific situations the sensory practitioner may want to be stricter and pick a smaller difference and in others a more lenient difference may be needed.) and the standard error associated with the study. The usual issue is that the practitioner does not know the standard error prior to the study, and so the tendency then is to "guess." Hough et al. (2006) evaluated 108 studies hedonic studies using consumers and found that the mean RMSL was 0.23 ± 0.037. (RMSL, root mean square divided by scale length. The root mean square of the error of a two-way ANOVA is equivalent to the standard error of the study. Since these authors were comparing 108 studies with some using the nine-point hedonic scale and others variations of this scale they had to standardize the root mean square.) They then calculated the number of consumers needed for a variety of scenarios with varying α, β, and size of the difference (Table 3.5). Similarly, Mammasse and Schlich (2014) did a series of resampling tests on the data from 7 hedonic tests and found that adequate panel sizes ranged from 20 to 150 depending on the complexity of the product space with more complex products (and most alcoholic beverages would fall in this category) needing larger numbers of consumers.

Table 3.5 Number of consumers needed for an acceptability test given an average value for RMSL

			β%		
RMSL	α%	d	**20**	**10**	**5**
0.23	10	0.2	17	23	29
	5	0.2	22	29	35
	1	0.2	32	40	48
	10	0.1	66	91	115
	5	0.1	84	112	138
	1	0.1	124	158	189
	10	0.05	262	363	459
	5	0.05	333	445	551
	1	0.05	495	631	755

α%, Probability of type I error; β%, probability of type II error; d, minimum difference in means required (scale 0–1); RMSL, root mean square error divided by scale length. Thus a nine-point hedonic scale is used and a 0.8 difference is required $d = 0.8/8 = 0.1$.

Adapted from Hough, G., Wakeling, I., Mucci, A., Chambers, E., Méndez Gallardo, I., Rangel Alves, L., 2006. Number of consumers necessary for sensory acceptability tests. Food Qual. Pref. 17, 522–526.

We would caution practitioners against asking consumers to indicate their hedonic responses, as well as asking them for intensity information on specific attributes associated with the beverage. As shown by Prescott et al. (2011) asking for attribute information places the consumers in an analytical mode and artificially decreases their hedonic ratings. Additionally, we also believe that not all consumers would understand the underlying concepts associated with the attribute, similarly leading to extremely noisy data—it is for this reason that we train panelists with reference standards, not something we can do with consumers. Therefore we strongly discourage this practice. As we will see further in this section there are statistical ways to integrate attribute information (derived by DA panelists) with consumer hedonic information. Similarly to preference ranking the samples used in a hedonic scaling study must be perceptibly different from one another for the consumers to indicate a difference in degree of liking (Guinard et al., 2001).

Hedonic scaling data are usually analyzed using parametric methods, despite the possibility that the data are more nearly ordinal than interval, because the large number of observations per product (usually about 100+) tends to approximate normality and makes the parametric statistical inferences more valid (Lim, 2011).

The data are usually evaluated by two-way ANOVA (main effects: consumers and products). If the products' hedonic scores vary significantly, then LSD is frequently used as the means separations technique for the products. However, since the data are wholly subjective, the standard deviations of the mean liking scores tend to be very high and it is entirely possible that the mean "is the result of diverging opinions" (Wajrock et al., 2008). Therefore, we strongly encourage practitioners to also look at the data from the individual consumer or from consumer cluster perspectives. This is usually done through the use of internal preference mapping (IPM) and/or external preference mapping. These two techniques "emphasize fundamentally different, but complementary perspectives on the data" (van Kleef et al., 2006).

IPM analyses create a two- or three-dimensional space based on the consumers' (or consumer clusters') liking scores for the products. The usual method is to do a PCA with the products as the rows and the individual consumers (or the mean cluster values) as the variables (Yenket et al., 2011). The resultant map can then be interpreted as all PCA maps are. If the sensory scientist is interested in which attributes, as used by trained panelists in a DA, describe the consumer derived space it is possible to regress the mean values for each attribute into the internal preference map space—this is called extended IPM (McEwan et al., 1998).

Conversely, in external preference mapping the two- or three-dimensional space is based on the objective attribute differences among the samples. The DA data are usually analyzed by PCA or CVA and then the consumer liking scores (or consumer hedonic mean cluster scores) are regressed into the space (Yenket et al., 2011). In this case and for the extended internal preference map, the originally created space is fixed with additional data projected into the space by regression. In the case of partial least squares (also known as projection to latent spaces or PLS) the two data sets (say X for the DA data and Y for the consumer data) are projected into the same space maximizing the covariance among them and thus maximizing the Y-data variance explained (MacKay, 2005).

van Kleef et al. (2006) showed that the maps obtained from extended IPM was more useful to marketing since these graphs emphasized consumer (dis)liking whereas the graphs from external preference mapping tended to be more useful for the food scientists attempting to optimize product attributes. A number of caveats need to be remembered when doing preference mapping: (1) the number of consumers fitting into the space can be as low as 50%—this is especially problematic when the study is performed on commercial products that do not greatly differ from one another; (2) drivers of preference do not always make "sense" (Guinard et al., 2001).

3.7 Conclusions

In summary, based on the goals of the experiment as discussed in Chapter 2, the sensory scientist will choose the appropriate method/s of analysis and statistical approaches. Frequently multiple approaches may be possible and the sensory scientist may be constrained in the choice of method by a number of factors, including the volume of sample available, the fatiguing nature of many alcoholic beverages, the number of samples to be tested, the changing nature of the sample over time, and so on. In the case study chapter we present actual experimental data for a variety of sensory methods and sample types, demonstrating applications, and where appropriate, the limitations of the methods for a range of alcoholic beverages.

References

Abdi, H., Williams, L.J., Valentin, D., 2013. Multiple factor analysis: principal component analysis for multitable and multiblock data sets. WIREs Comp. Stats..

Abdi, H., Valentin, D., Chollet, S., Chrea, C., 2007. Analyzing assessors and products in sorting tasks: DISTATIS, theory and applications. Food Qual. Pref. 18, 627–640.

Angulo, O., O'Mahony, M.O., 2005. The paired preference test and the "No Preference" option: was Odesky correct? Food Qual. Pref. 16, 425–434.

Arnold, G.M., Gower, J.C., Gardner-Lubbe, S., le Roux, N.J., 2007. Biplots of free-choice profile data in generalized orthogonal Procrustes analysis. J. R. Stat. Soc. C Appl. Stat. 56, 445–458.

Barcenas, P., Pérez Elortondo, F.J., Albisu, M., 2004. Projective mapping in sensory analysis of ewes milk cheeses: a study on consumers and trained panel performance. Food Res. Int. 37, 723–729.

Bartoshuk, L.M., Cartalanotto, J.A., Hoffman, H.J., Logan, H.L., Snyder, D.J., 2012. Taste damage (otitis media, tonsillectomy and head and neck cancer) can intensify oral sensations. Physiol. Behav. 107, 516–526.

Bausell, R.B., Li, Y.-F., 2002. Power Analysis for Experimental Research. Cambridge University Press, Cambridge, UK.

Bi, J., O'Mahony, M., 2013. Variance of the d' for the tetrad test and comparisons with other forced-choice methods. J. Sens. Stud., 91–101.

Blancher, G., Chollet, S., Kesteloot, R., Nguyen Hoang, D., Cuvelier, G., Sieffermann, J.-M., 2007. French and Vietnamese: how do they describe texture characteristics of the same food? A case study with jellies. Food Qual. Pref. 18, 560–575.

Bovell-Benjamin, A.C., Allen, L.H., Guinard, J.X., 1999. Toddlers' acceptance of whole maize porridge fortified with ferrous biglycinate. Food Qual. Pref. 10, 123–128.

Brockhoff, P., Christensen, R., 2010. Thurstonian models for sensory discrimination tests as general linear models. Food Qual. Pref. 21, 330–338.

Bruzzone, F., Ares, G., Giménez, A., 2013. Temporal aspects of yoghurt texture perception. Int. Dairy J. 29, 124–134.

Campo, E., Ballester, J., Langlois, J., Dacremont, C., Valentin, E., 2010. Comparison of conventional descriptive analysis and a citation frequency-based descriptive method for odor profiling: an application to Burgundy Pinot noir wines. Food Qual. Pref. 21, 44–55.

Chapman, K.W., Grace-martin, K., Lawless, H.T., 2006. Expectations and stability of preference choice. J. Sens. Stud. 21, 441–455.

Christensen, R.H.B., Brockhoff, P.B., 2015. sensR—an R-package for sensory discrimination. R package version 1.4-5. Available from: http://www.cran.r-project.org/package=sensR/

Clark, C.C., Lawless, H.T., 1994. Limiting response alternatives in time–intensity scaling: an examination of the halo-dumping effect. Chem. Senses 19, 583–594.

Cristovam, E., Paterson, A., Piggott, J.R., 2000. Differentiation of port wines by appearance using a sensory panel: comparing free choice and conventional profiling. Eur. Food Res. Technol. 211, 65–71.

Dairou, V., Sieffermann, J.-M., 2002. A comparison of 14 jams characterized by conventional profile and a quick original method, the flash profile. J. Food Sci. 67, 826–834.

Dehlholm, C., Brockhoff, P.B., Meinert, L., Aaslyng, M., Bredie, W.L.P., 2012. Rapid descriptive methods—comparison of free multiple sorting, partial napping, napping, flash profiling and conventional profiling. Food Qual. Pref. 26, 267–277.

Delarue, J., Lawlor, B., Rogeaux, M., 2015. Rapid Sensory Profiling Techniques: Applications in New Product Development and Consumer Research. Woodhead Publishing Series in Food Science, Technology and Nutrition. Elsevier, New York.

Déléris, I., Saint-Eve, A., Guo, Y., Lieben, P., Cypriani, M.-L., Jacquet, N., Brunerie, P., Souchon, I., 2011. Impact of swallowing on the dynamics of aroma release and perception during the consumption of alcoholic beverages. Chem. Senses. 36, 701–713.

Di Monaco, R., Su, C., Masi, P., Cavella, S., 2014. Temporal dominance of sensations: a review. Trends Food Sci. Technol. 38, 104–112.

Dinella, C., Masi, C., Naes, T., Monteleone, E., 2013. A new approach in TDS data analysis: a case study on sweetened coffee. Food Quality Pref. 30, 33–46.

Donaldson, B.A., Bamforth, C.W., Heymann, H., 2012. Sensory descriptive analysis and free-choice profiling of thirteen hop varieties as whole cones and after dry hopping of beer. J. Am. Soc. Brew. Chem. 70, 176–181.

Duizer, L.M., Bloom, K., Findlay, C.J., 1997. Dual-attribute time intensity sensory evaluation: a new method for temporal measurement of sensory perceptions. Food Qual. Pref. 8, 261–269.

Duizer, L.M., Bloom, K., Findlay, C.J., 1996. Dual-attribute time–intensity measurement of sweetness and peppermint perception of chewing gum. J. Food Sci. 61, 636–638.

Elmore, J.R., Heymann, H., 1999. Perceptual maps of photographs of carbonated beverages created by traditional and free-choice profiling. Food Qual. Pref. 10, 219–227.

Ennis, J.M., Rousseau, B., Ennis, D.M., 2014. Sensory difference tests as measurement instruments: a review of recent advances. J. Sens. Stud. 29, 89–102.

Ennis, D.M., 1993. The power of sensory discrimination methods. J. Sens. Stud. 8, 353–370.

Escofier, B., Pagès, J., 1990. Multiple factor analysis. Comp. Stat. Data Anal. 18, 121–140.

Gacula, M.C., Rutenbeck, S., 2006. Sample size in consumer tests and descriptive analysis. J. Sens. Stud. 21, 129–145.

Gacula, M.C., Singh, J., 1984. Statistical Methods in Food and Consumer Research. Academic Press, Inc., Orlando, FL, 84–93.

Garcia, K., Ennis, J.M., Prinyawiwatkul, W., 2012. A large-scale experimental comparison of the tetrad and triangle tests in children. J. Sens. Stud. 27, 217–222.

Gilbert, J., Heymann, H., 1995. Comparison of four sensory methodologies as alternatives to descriptive analysis for the evaluation of apple essence aroma. N. Z. Food Technol. 24 (4), 28–32.

Goupil de Bouille, A., Pineau, N., Meyners, M., Martin, N., Schlich, P., 2010. How do panelists use the list of attributes during a temporal dominance of sensations experiment? In: Proceedings of the 11th European Symposium on Statistical Methods for the Food industry (AgroStat), Benevento, Italy, February 23–26, 179–186.

Guinard, J.-X., Uotani, B., Schlich, P., 2001. Internal and external mapping of preferences for commercial lager beers: comparison of hedonic ratings by consumers blind versus with knowledge of brand and price. Food Qual. Pref. 12, 243–255.

Guinard, J.-X., Zoumas-Morse, C., Dietz, J., Goldberg, S., Holz, M., Heck, E., Amoros, A., 1996. Does consumption of beer, alcohol, and bitter substances affect bitterness perception? Physiol. Behav. 59, 625–631.

Hein, K.A., Hamid, N., Jaeger, S.R., Delahunty, C.M., 2012. Effects of evoked consumption contexts on hedonic ratings: a case study with two fruit beverages. Food Qual. Pref. 26, 35–44.

Heymann, H., King, E.S., Hopfer, H., 2014. Classical descriptive analysis. In: Ares, G., Varela, P. (Eds.), Novel Techniques in Sensory Characterization and Consumer Profiling. CRC Press, New York, pp. 9–40.

Heymann, H., Noble, A.C., 1989. Comparison of canonical variate and principal component analyses of wine descriptive analysis data. J. Food Sci. 54, 1355–1358.

Hopfer, H., Heymann, H., 2014. Judging wine quality: do we need experts, consumers or trained panelists? Food Qual. Pref. 32, 221–233.

Hough, G., Wakeling, I., Mucci, A., Chambers, E., Méndez Gallardo, I., Rangel Alves, I., 2006. Number of consumers necessary for sensory acceptability tests. Food Qual. Pref. 17, 522–526.

Ishii, R., Kawaguchi, H., O'Mahony, M., Rousseau, B., 2013. Relating consumer and trained panels' discriminative sensitivities using vanilla flavored ice cream as a medium. Food Qual. Pref. 18, 89–96.

Jones, L.V., Peryam, D.R., Thurstone, L.L., 1955. Development of a scale for measuring soldiers' food preferences. Food Res. 20, 512–520.

Kalva, J.J., Sims, C.A., Puentes, L.A., Snyder, D.J., Bartoshuk, L.M., 2014. Comparison of the hedonic general labeled magnitude scale with the hedonic 9-point scale. J. Food Sci. 79, S238–S245.

Kobue-Lekalake, R.I., Taylor, J.R.N., de Kock, H., 2011. Application of the dual attribute time–intensity (DATI) sensory method to the temporal measurement of bitterness and astringency in sorghum. Int. J. Food Sci. Technol. 47, 459–466.

Köster, E.P., Couronne, T., Léon, F., Lévy, C., Marcelino, A.S., 2003. Repeatability in hedonic sensory measurement: a conceptual exploration. Food Qual. Pref. 14, 165–176.

Kozak, M., Cliff, M.A., 2013. Systematic comparison of hedonic ranking and rating methods demonstrates few practical differences. J. Food Sci. 78, S1257–S1263.

Labbe, D., Schlich, P., Pineau, N., Gilbert, F., Martin, N., 2009. Temporal dominance of sensations and sensory profiling: a comparative study. Food Qual. Pref. 20, 216–221.

Lahne, J., Collins, T.S., Heymann, H., 2016. Replication improves sorting-task results analyzed by DISATIS in a consumer study of American Bourbon and Rye whiskeys. J. Food Sci..

Lawless, H.T., 2013a. Thurstonian models for discrimination and preference. John Wiley and Sons, New York, 71–98 (Chapter 4).

Lawless, H.T., 2013b. Progress in discrimination testing. John Wiley and Sons, New York, 99-123.

Lawless, H.T., 2013c. Time–intensity modeling. John Wiley and Sons, New York, 240–256 (Chapter 11).

Lawless, H.T., Glatter, S., 1990. Consistency of multidimensional scaling models derived from odor sorting. J. Sens. Stud. 5, 217–230.

Lawless, H.T., Heymann, H., 2010. Sensory Evaluation of Food: Principles and Practices, second ed. Springer, New York, NY.

Lawless, H.T., Popper, R., Kroll, B.J., 2009. A comparison of the labeled magnitude (LAM) scale, an 11-point category scale and the traditional 9-point hedonic scale. Food Qual. Pref. 21, 4–12.

Lê, S., Husson, F., 2008. SensoMineR: a package for sensory data analysis. J. Sens. Stud. 23, 14–25.

Lê, S., Pagès, J., Husson, F., 2008. Methodology for the comparison of sensory profiles provided by several panels: application to a cross-cultural study. Food Qual. Pref. 19, 179–184.

Lea, P., Næs, T., Rødbotten, M., 1997. Analysis of variance for sensory data. John Wiley and Sons, New York.

Leach, E.J., Noble, A.C., 1986. Comparison of bitterness of caffeine and quinine by a time–intensity procedure. Chem. Senses 11, 339–345.

Lim, J., 2011. Hedonic scaling: a review of methods and theory. Food Qual. Pref. 22, 733–747.

Lim, J., Wood, A., Green, B.G., 2009. Derivation and evaluation of a labeled hedonic scale. Chem. Senses 34, 739–751.

MacKay, D., 2005. Chemometrics, econometrics, psychometrics—how best to handle hedonics? Food Qual. Pref. 17, 529–535.

Mammasse, N., Schlich, P., 2014. Adequate numbers of consumers in a liking test. Insights from resampling in seven studies. Food Qual. Pref. 31, 124–128.

Marchisano, C., Lim, J., Cho, H.S., Suh, D.S., Jeon, S.Y., Kim, K.O., O'Mahony, M., 2003. Consumers report preferences when they should not: a cross-cultural study. J. Sens. Stud. 18, 487–516.

Marley, A.A.J., Louviere, J.J., 2005. Some probabilistic models of bets, worst and best-worst choices. J. Math. Psychol. 49, 464–480.
McEwan, J.A., Earthy, P.J., Ducher, C., 1998. Preference mapping: a review. Review No. 6, Project No. 29742, Campden and Chorleywood Food Research Association, UK.
Meilgaard, M.C., Carr, B.T., Civille, G.V., 2007. Sensory evaluation techniques, fourth ed. Taylor and Francis/CRC Press, Boca Raton.
Meillon, S., Urbano, C., Schlich, P., 2009. Contribution of the temporal dominance of sensations (TDS) method to the sensory description of subtle differences in partially dealcoholized red wines. Food Qual. Pref. 20, 490–499.
Meyners, M., 2010. On the design, analysis, and interpretation of temporal dominance of sensations data. In: Proceedings of the 11th European Symposium on Statistical Methods for the Food industry (AgroStat), Benevento, Italy, February 23–26, 45–53.
Meyners, M., 2007. Easy and powerful analysis of replicated paired preference tests using the χ^2 test. Food Qual. Pref. 18, 938–948.
Mielby, L.H., Edelenbos, M., Thybo, A.K., 2012. Comparison of rating, best-worst scaling, and adolescents' real choices of snacks. Food Qual. Pref. 25, 140–147.
Mielby, L.H., Hopfer, H., Jensen, S., Thybo, A.K., Heymann, H., 2014. Comparison of descriptive analysis, projective mapping and sorting performed on pictures of fruit and vegetable mixes. Food Qual. Pref. 35, 86–94.
Moskowitz, H.R., Sidel, J.L., 1971. Magnitude and hedonic scales of food acceptability. J Food Sci. 36, 677–680.
Mueller, S., Francis, I.L., Lockshin, L., 2009. Comparison of best-worst and hedonic scaling for the measurement of consumer wine preferences. Aust. J. Grape Wine Res. 15, 205–215.
Næs, T., Brockhoff, P.B., Tomic, O., 2010. Statistics for Sensory and Consumer Science. John Wiley and Sons, New York.
Newell, G.J., MacFarlane, J.D., 1987. Expanded tables for multiple comparison procedures in the analysis of ranked data. J. Food Sci. 52, 1721–1725.
Odesky, S.H., 1967. Handling the neutral vote in paired comparison product testing. J. Mark. Res. 4, 199–201.
Pagès, J., 2005. Collection and analysis of perceived product inter-distances using multiple factor analysis: application to the study of 10 white wines from the Loire Valley. Food Qual. Pref. 16, 642–649.
Pagès, J., Husson, F., 2005. Multiple factor analysis with confidence ellipses: a methodology to study the relationships between sensory and instrumental data. J. Chemometrics 19, 138–144.
Peltier, C., Visalli, M., Schlich, P., 2015. Comparison of canonical variate analysis and principal component analysis on 422 descriptive sensory studies. Food Qual. Pref. 40, 326–333.
Perrin, L., Symoneaux, R., Maître, I., Asselin, C., Jourjon, F., Pagès, J., 2007. Comparison of conventional profiling by a trained tasting panel and free profiling by wine professionals. Am. J. Enol. Vitic. 58, 508–517.
Perrin, L., Pagès, J., 2009. Construction of a product space from the ultra-flash profiling method: application to ten Loire red wines. J. Sens. Stud. 24, 372–395.
Peryam, D.R., Girardot, N.F., 1952. Advanced taste test method. Food Eng. 24, 58–61, 194.
Peryam, D.R., Pilgrim, F.J., 1957. Hedonic scale method of measuring food preferences. Food Technol., 9–14.
Pickering, G.J., Heatherbell, D.A., Vanhanen, L.P., Barnes, M.F., 1998. The effect of ethanol concentration on the temporal perception of viscosity and density in white wine. Am. J. Enol. Vitic. 49, 306–318.
Piggott, J.R., Watson, M.P., 1992. A comparison of free-choice profiling and the repertory grid method in the flavor profiling of cider. J. Sens. Stud. 7, 133–145.
Pineau, N., Schlich, P., Cordelle, S., Mathonnière, C., Issanchou, S., Imbert, A., et al., 2009. Temporal dominance of sensations: construction of the TDS curves and comparison with time–intensity. Food Qual. Pref. 20, 450–455.

Pineau, N., Goupil de Bouille, A., Lepage, M., Lenfant, F., Schlich, P., Martin, N., et al., 2012. Temporal dominance of sensations: what is a good attribute list? Food Qual. Pref. 26, 159–165.

Popper, R., Heymann, H., 1996. Analyzing differences among products and panelists by multidimensional scaling. In: Naes, T., Risvik, E. (Eds.), Multivariate Analysis of Data in Sensory Science. Elsevier Science B.V., pp. 159–184.

Prescott, J., Lee, S.M., Kim, K.-O., 2011. Analytic approaches to evaluation modify hedonic responses. Food Qual. Pref. 22, 391–393.

Risvik, E., McEwan, J.A., Colwill, J.S., Rogers, R., Lyon, D.H., 1994. Projective mapping: a tool for sensory analysis and consumer research. Food Qual. Pref. 5, 263–269.

Risvik, E., McEwan, J.A., Rødbotten, M., 1997. Evaluation of sensory profiling and projective mapping. Food Qual. Pref. 8, 63–71.

Roessler, E.B., Pangborn, R.M., Sidel, J.L., Stone, H., 1978. Expanded statistical tables for estimating significance in paired-preference, paired-difference, duo–trio and triangle tests. J. Food Sci. 43, 940–947.

Savidan, C.H., Morris, C., 2015. Panelists' performances and strategies in paper-based and computer-based projective mapping. J. Sens. Stud. 30, 145–155.

Schlich, P., 1993. Risk tables for discrimination tests. Food Qual. Pref. 4, 141–151.

Schutz, H.G., Cardello, A.V., 2001. A labeled affective magnitude (LAM) scale for assessing food liking/disliking. J. Sens. Stud. 16, 117–159.

Sheskin, D.J., 2007. Handbook of parametric and non-parametric statistical procedures. Chapman and Hall/CRC, Boca Raton, FL, 1075–1088.

Sokolowsky, M., Fischer, U., 2012. Evaluation of bitterness in white wine applying descriptive analysis, time–intensity analysis, and temporal dominance of sensations analysis. Anal. Chim. Acta 732, 46–52.

Stone, H., Bleibaum, R., Thomas, H.A., 2012. Sensory evaluation practices, fourth ed. Academic Press, London, England.

Swaney-Stueve, M., Heymann, H., 2002. A comparison between paper and computerized ballots and a study of simulated substitution between the two ballots used in descriptive analysis. J. Sens. Stud. 17, 527–537.

Tang, C., Heymann, H., 2002. Multidimensional sorting, similarity scaling and free-choice profiling of grape jellies. J. Sens. Stud. 17, 493–509.

Tarea, S., Cuvelier, G., Sieffermann, J.M., 2007. Sensory evaluation of the texture of 49 commercial apple and pear purees. J. Food Qual. 30, 1121–1131.

Teillet, E., Schlich, P., Urbano, C., Cordelle, S., Guichard, E., 2010. Sensory methodologies and the taste of water. Food Qual. Pref. 21, 967–976.

Thullier, B., Valentin, D., Marchal, R., Dacremont, C., 2015. Pivot© profile: a new descriptive method based on free description. Food Qual. Pref. 42, 66–77.

Tomic, O., 2013. Differences between generalised procrustes analysis and multiple factor analysis in case of projective mapping, master's thesis. Available from: https://brage.bibsys.no/xmlui/bitstream/id/185096/MasterThesis_OliverTomic.pdf

Valentin, D., Chollet, S., Lelièvre, M., Abdi, H., 2012. Quick and dirty but still pretty good: a review of new descriptive methods in food science. Int. J. Food Sci. Technol. 47, 1563–1578.

van Kleef, E., van Trijp, H.C.M., Luning, P., 2006. Internal versus external preference analysis: an exploratory study on end-user evaluation. Food Qual. Pref. 17, 387–399.

Varela, P., Ares, G., 2014. Novel Techniques in Sensory Characterization and Consumer Profiling. CRC Press, New York.

Vázquez-Araújo, L., Parker, D., Woods, E., 2013. Comparison of temporal-sensory methods for beer flavor evaluation. J. Sens. Sci. 28, 387–395.

Wajrock, S., Antille, N., Rytz, A., Pineau, N., Hager, C., 2008. Partitioning methods outperform hierarchical methods for clustering consumers in preference mapping. Food Qual. Pref. 18, 662–669.

Williams, A.A., Langron, S.P., 1984. The use of free-choice profiling for the evaluation of commercial ports. J. Sci. Food Agric. 35, 558–568.
Yenket, R., Chambers, E., Adhikari, K., 2011. A comparison of seven preference mapping techniques using four software programs. J. Sens. Stud. 26, 135–150.

In this chapter

Uses of chemical testing in the alcoholic beverage industry

4.1 Introduction

Chemical analyses in the alcoholic beverage industry provide a fundamental basis for many of the day-to-day decisions made by winemakers, brewers, and distillery managers. Frequently, sensory assessment on a formal or informal basis can inform production decisions, however, chemical analysis can also be crucial for preventing and solving problems in a timely manner and for ensuring consistent and stable products. Further, many chemical analyses are required for tax or legal purposes. The choice of the analytical procedures will depend on the uses of the analytical information, as well as the desired turnaround times for the information and the performance measures of the available methods.

In this chapter we will review some of the basic uses and purposes of chemical testing for the alcoholic beverage industry. The focus here is on analytes that exemplify specific purposes; the analyses discussed are not meant to provide a comprehensive listing of all analytes that are routinely measured in alcoholic beverages. In subsequent chapters we will focus on some of the common analytes measured, the methodologies for their analysis, and the performance measures required for a variety of purposes.

4.2 Confirm composition of starting materials or ingredients

In the initial production stages, chemical analyses are used to confirm that ingredient properties are consistent with the specifications and/or requirements of the product being produced. Conversion of sugars by yeast is a defining feature of fermented beverage production, therefore, one of the main components affecting final product composition and quality is the total or fermentable sugar level of the starting material. Based on biochemical conversion of the simple

Sensory and Instrumental Evaluation of Alcoholic Beverages. http://dx.doi.org/10.1016/B978-0-12-802727-1.00004-1

sugar glucose to ethanol, conversion rates will be 51% on a mass basis (Eq. 4.1). Actual conversion rates may vary depending on fermentation conditions, the actual sugars that are present, and whether the sugars are completely utilized by the yeast. The concentrations of residual, unfermented sugars will influence the perceived sweetness of the final product as well as microbial stability.

$$\begin{aligned} \text{Glucose} &\rightarrow \text{Ethanol} + \text{Carbon dioxide} \\ C_6H_{12}O_6 &\rightarrow 2C_2H_5OH + 2CO_2 \end{aligned} \tag{4.1}$$

In the wine industry, grape composition is monitored throughout the growing season and harvest decisions are typically made based on an analysis of the total sugar level. Grape berries accumulate sugar during maturation with a rapid accumulation occurring at the onset of grape berry ripening, also called véraison (Fig. 4.1). Generally, the increase in sugar concentration will continue until harvest, although the rate of increase slows and is also often impacted by berry dehydration if harvest is delayed. Sugar levels at harvest typically range from 18% to 19% (by weight) to as much as 39% (by weight) or more.

For beer and malt beverage production, the total fermentable sugars in the malt are monitored. During beer making, malted grains are crushed and milled before mixing with hot water to gelatinize the starch, a complex carbohydrate. Enzymes from the grain then begin to breakdown the starch into the fermentable sugars (Bamforth, 2000). Total sugar levels in the mash are a function of the grain type, individual cultivars, maturity, and sprouting conditions, and the addition of other grain adjuncts (Russell and Stewart, 2003; Priest and Stewart, 2006). Typically,

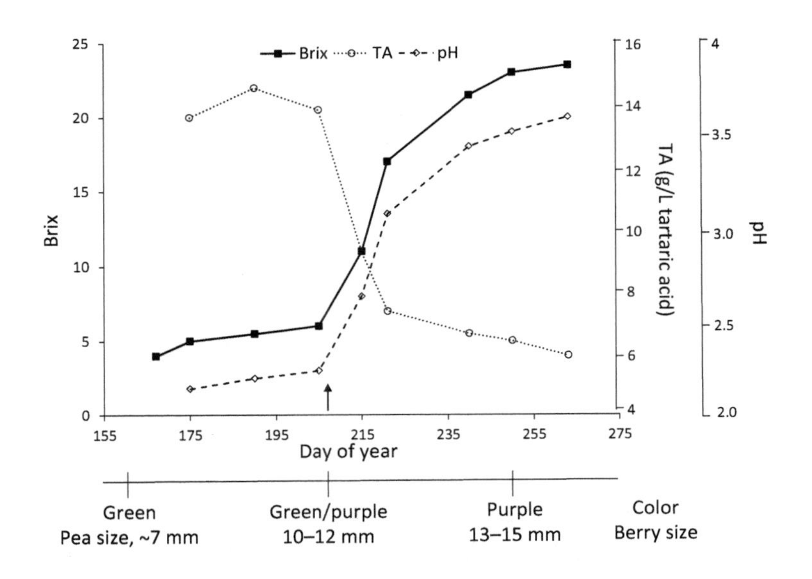

Figure 4.1
Changes in Brix, titratable acidity (TA), pH, color, and size of grape berries during maturation (northern hemisphere)
All values shown are approximate and will vary with grape variety, growing region, viticultural practices, and year. The *arrow* on the *x*-axis notes the onset of ripening, véraison.

total fermentable sugar levels in malt will be ~80% but can vary with mashing conditions and temperature and desired beer style (Muller, 1991, 2000).

As total sugar levels influence alcohol yield, grapes and malt are often priced based on their sugar level and producers may contract with growers or suppliers to purchase grapes or malt at defined sugar levels. Producers may make fermentation decisions based on initial sugar levels in order to maximize the economic efficiencies of ethanol conversion. Especially if the fermented product will then be used for subsequent distillation, the ethanol yield from the primary fermentation becomes critical.

In addition to the sugars needed for energy, yeast also need nutrients, such as nitrogen sources (ammonia and/or amino acids) and vitamins to grow and reproduce. Insufficient nutrients can result in fermentations that proceed very slowly or even stop before all of the sugar is consumed. There are numerous methods for measuring nitrogen and nutrient levels and nutrients may be added to the fermentation mixtures if the levels are too low.

The pH and acidity [titratable acidity (TA)] of starting materials may be monitored since they can significantly impact chemical and microbial reactions, as well as the sensory properties, including the color, of the final beverage. During grape maturation, berry pH increases and TA decreases (Fig. 4.1), resulting in a decrease in perceived sourness or tartness. A balance in the concentrations of sugars and acids can be important to ensure that the final product is not "cloying" or too tart. pH and acidity may also impact chemical and biochemical processes and microbial stability. The pH and acidity of the starting grape juice or must do not typically change significantly during wine fermentation, so the starting pH and TA will determine the final pH and TA unless acid additions occur during processing.

Water is an important ingredient in many alcoholic beverages, particularly beer. Therefore, water quality during brewing is particularly important as it can influence activity of enzymes that convert complex carbohydrates to fermentable sugars and formation of hazes and precipitates during beer making (Bamforth, 2000). As a result, water hardness and specific elements (e.g., calcium, magnesium) may be monitored (Bamforth, 2016).

Finally, the color and the composition of aroma or other taste active compounds may be measured. For pigmented grapes, the color increases during maturation (Fig. 4.1) and may impact harvest decisions. The malt roasting process will determine the color of the malt and is an important factor determining the color of the final beer or malt beverage (Iserentant, 1995). Taste active compounds, such as isoalpha acids in hops, which contribute to bitterness, are an important measure when determining the hops cultivars to use for a specific beer style. Aroma compounds which contribute to green, grassy, or bell pepper aromas typically decrease during grape berry maturation and so may be monitored when making harvest decisions. Other aroma compounds may increase during maturation, contributing to fruity and/or varietal aromas (Canuti et al., 2009). Increasingly, wineries may monitor these aroma compounds as a full-spectrum "quality management" program when purchasing grapes and working with growers/viticulturists to manage vineyard practices and to make harvest decisions (Cleary et al., 2015). Off-aromas in the water used during processing should also be avoided. If off-aromas are present in the water supply, additional water treatments may be needed prior to use (Bamforth, 2016).

4.3 Monitor production processes

Numerous analytes may be monitored during production of fermented beverages. The most fundamental process for fermented beverages is the utilization of sugar and production of alcohol by yeast; as a result, sugar and/or alcohol levels are often monitored during fermentation in order to detect potential problems. For example, if the fermentation rate slows or stalls before the sugar is completely consumed, winemakers can make decisions about how to "restart" the fermentation (i.e., add nutrients, add more yeast, etc.) or they can choose to use the wine for blending or in a different product stream. Similarly, brewers will monitor sugar consumption to ensure that the fermentation is proceeding as expected.

During brewing, the pH of the malt and water (termed, the mash, when mixed together) can impact enzyme activity. Brewers typically aim for a mash pH of 5.0–5.5 for optimal enzyme activity during the brewing process and when the pH is not in the optimal range salts or buffers may be added.

During distillation, the alcohol content is monitored in order to determine the cut-point. At early stages of distillation (the "heads fraction"), the alcohol composition will be low (~35% by volume) and contain relatively high concentrations of volatile compounds, such as aldehydes and methanol which are toxic. After most of the ethanol has distilled over (the "heart fraction"), higher alcohols or fusel oils will begin to distill and contribute fusel-like aromas (the "tail fraction"). In addition, the concentration of water in the distillate increases.

4.4 Monitor and ensure stability during storage (chemical and microbial)

During storage, the chemical composition can have a significant influence on the chemical and microbial stability of alcoholic beverages. Residual sugars, ethanol, pH, and sulfur dioxide (SO_2) levels can all interact to influence overall stability. Beverages with pH > 4.0, alcohol concentrations <7–10%, and/or free SO_2 levels less than ~10–20 mg/L will be susceptible to microbial growth which may result in formation of off-flavors and CO_2 during storage (Ough and Amerine, 1988).

If residual fermentable sugars remain, microorganisms, such as the yeast *Brettanomyces*, can grow during storage, resulting in formation of off-flavors and/or CO_2 production. The aroma compounds, 4-ethylphenol and 4-ethylguaiacol, which contribute to the phenolic/band-aid-like off-aroma of *Brettanomyces* contamination, can be monitored, usually by gas chromatographic methods to determine the extent of off-flavor production. Similarly, gas chromatographic analysis of *trans*-2-nonenal and 3-methylbut-2-ene-1-thiol which contribute to oxidized/cardboard-like and skunky aromas, respectively, may be monitored in beer. Oxidative reactions and photodegradation of isoalpha acids lead to formation of these off-aromas during storage.

4.5 Consistency and quality assurance/quality control

As more is learned about the chemical composition of fermented beverages, it is becoming increasingly possible to use one or more chemical components to ensure product consistency and uniformity among batches and over time. For example, beer manufacturers may monitor dimethyl sulfide (DMS) levels or concentrations of isoalpha acids in a specific beer style or product line (Scarlata and Ebeler, 1999; Miracle et al., 2005).

This concept is being extended to the analysis of complex chemical profiles (including volatiles, nonvolatiles, and elemental composition) to monitor authenticity of fermented beverages (Versari et al., 2014). For example, recent studies have shown that the chemical composition of Malbec wines differs for grapes grown in Argentina compared to California (King et al., 2014; Nelson et al., 2015). Similarly, whiskey styles (American, Tennessee, bourbon, rye) as well as whiskeys made by different producers and whiskeys of different ages can be differentiated based on the complex chemical profiles (Collins et al., 2014). This information may make it possible to monitor labeling information with respect to geographic origin, age, and starting material (grape variety, grain type, etc.).

4.6 Ensure compliance with tax, legal, and safety regulations

In the United States, alcoholic beverages are taxed based on the alcohol content, therefore alcohol measurements are one of the most important analyses performed. Levels of CO_2 are also used to distinguish tax levels for sparkling and still wines. Maximum levels of volatile acidity are strictly regulated since high levels can be an indicator of unsound winemaking practices. Additional compounds may be monitored for health or safety reasons. For example, SO_2 levels are regulated to protect subpopulations that are highly allergic to sulfites. Recently a number of additional additives that may be used to treat alcoholic beverages have come under scrutiny for the potential of residues to trigger allergies in consumers. Food allergens, such as milk, fish, and proteins have received much attention (Peñas et al., 2015). Table 4.1 summarizes the regulatory requirements for selected analytes in alcoholic beverages.

The main regulatory agency in the United States governing alcoholic beverages is the Alcohol and Tobacco Tax and Trade Bureau (www.ttb.gov). Information on standards of identity, labeling requirements, and a full listing of materials that may be added to alcoholic beverages in the United States can be found in the Code of Federal Regulations (CFR) Title 27 (Alcohol, Tobacco Products and Firearms) with individual regulations governing distilled spirits (Part 5), wine (Parts 4 and 24), and beer (Parts 7 and 25). In other countries legal limits may be different and regulatory agencies should be consulted, for example, International Organisation of Vine and Wine (OIV) in the European Union (http://www.oiv.int/oiv/cms/index?lang=en) and the Liquor Control Boards of Canada (http://www.canadianvintners.com/canadas-industry/liquor-boards-of-canada/).

Table 4.1 Legal requirements for selected analytes in alcoholic beverages for the United States

Analyte	Legal requirements	Product
Ethanol	$<$14.0% (v/v)	Table wine
Ethanol	$>$14.0 and $<$24% (v/v)	Dessert wine
Ethanol	Variable	Distilled spirits—see specific standards of identity (US CFR 27.5)
Ethanol	$<$0.5% (v/v)	Malt beverage/cereal beverage/near beer
CO_2	$<$3.92 g/L	Still wine
CO_2	$>$3.92 g/L	Sparkling wine
Volatile acidity	$<$1.2 g/L (as acetic acid) $<$1.4 g/L (as acetic acid)	White wine Red wine
SO_2	$>$10 mg/L total SO_2	All beverages, label warning required
SO_2	Maximum 350 mg/L total SO_2	All wines
Food allergens (e.g., milk, fish, egg, proteins, wheat, and so on)	Voluntary	All alcoholic beverages
Pesticide residues	MRL vary depending on pesticide and product	Grapes, grain, hops (European Commission No. 396/2005; US CFR 40.180)

MRL, Maximum residue limits.

In some cases, materials may be added during production of alcoholic beverages and the regulations limit the maximum residue levels that can remain in the final product. An example is copper, which may be added during wine production to minimize off-aromas from reduced sulfur compounds, such as hydrogen sulfide and methane thiol—US regulations allow up to 6 mg/L copper to wines, however, the residue levels after bottling cannot exceed 0.5 mg/L (as copper) (US CFR, Title 27, Part 24, 24.246). Pesticides provide another example. When approved for use on foods and commodities, the US Environmental Protection Agency (US EPA) sets limits on the maximum amount of a residue than can remain on the food. The FDA and USDA are authorized to monitor pesticide residues in foods that may be sold in the United States and the Code of Federal Regulations provides information on maximum residue levels that have been approved (CFR, Title 40, Part 180: Tolerances and exemptions from tolerances for pesticide chemicals in food). The US Alcohol, Tobacco Tax and Trade Bureau enforces residue limits under provisions of these regulations.

Products may also be monitored to ensure that they comply with label designations for country/region of origin, grape variety and/or source of fermentable sugars, and product age. Dilution with water, addition of ethanol from other sources, and use of colorants or flavors may be regulated and accidental or intentional misrepresentation of a product can result in legal consequences for producers. Regulations and requirements vary by country and the appropriate regulatory agencies should be carefully consulted to ensure that approved processes and ingredients, labeling requirements, and record-keeping are followed with respect to authentication and adulteration of alcoholic beverages.

4.7 Flavor and off-flavor assessment

Volatile aroma and nonvolatile taste and/or mouthfeel components may be monitored in order to assess potential flavor quality or flavor attributes of alcoholic beverages. In a relatively small number of cases, individual compounds may be associated with specific aroma qualities, for example, 2-isobutyl-3-methoxypyrazine, which has a bell pepper aroma contributes to the varietal character of some grapes and wines, as noted previously. Dimethyl sulfide levels in beer also are closely monitored since low levels contribute to characteristic aroma profiles while high levels give beer an undesirable cooked vegetable or cabbage-like odor (Scarlata and Ebeler, 1999). The compound 2, 4, 6-trichloroanisole has a distinct musty off-aroma and analysis of this compound is standard for screening corks and containers used to store wine. In most cases however, flavor perception is the result of multiple compounds interacting together and it is difficult to predict flavor qualities from specific compounds alone. Analysis of compounds contributing to aroma, taste, and mouthfeel will be discussed in detail in Chapters 5 and 6.

4.8 Developing process analytical charts

In the previous sections we reviewed some general purposes for performing chemical testing and briefly discussed how the information may be used. In any given production setting it is always important to ask why analyses are done and how the information from the analysis will be used to inform the production process—if the answers to the questions are not apparent, then it is important to ask whether the analysis should continue to be performed. A valuable approach for determining the critical processing points that impact any aspect of food quality maintenance is to use the basic framework of Hazard Analysis and Critical Control Points (HACCP) assessment (http://www.fda.gov/Food/GuidanceRegulation/HACCP/ucm2006801.htm#execsum; accessed August 7, 2015). Although initially devised for identifying critical points in food processing that can result in concerns with respect to microbial food safety, the HACCP process can be used to evaluate every step in the production of a food/beverage, determine the points where critical control and/or processing decisions are made, and then evaluate and define the variables needed to make and implement the controls/decisions.

When a HACCP analysis is applied to the use and application of analytical measurements, a process analytical chart is obtained. Starting with a flow chart of the entire production process, the critical control/decision-making points and information on the analyses needed to make informed decisions are identified, as well as the range in analyte concentrations needed to affect a decision. The process analytical chart would specify the timeline for when information is needed to efficiently and effectively make the process decision, what method will be used to measure the analyte/s of interest, and the degree of accuracy/precision needed in the analysis.

For example, a critical point in winemaking is the harvest decision—are the grapes ready to harvest or not. When making a red table wine, a winery might determine that a total sugar level of 23–24% (Brix) is optimal for that particular wine brand/style, therefore, once the grapes reach this sugar level, the decision is made to initiate the grape harvest in a given vineyard. The winemaker determines that rapid turnaround methods are needed to determine the total sugar level at this point in the production cycle (delays of 24 h could result in significant changes in the berry sugar levels), and the accuracy of the method necessary to make the decision would be, for example, ±1 Brix. The process chart would also specify any specific sampling methods appropriate to ensure that a representative sample has been obtained prior to the analysis.

Using this approach, many analyses could be identified that "could" be used to make decisions, however, only those analyses that truly inform the decision-making process should be performed. For example, if a winery does not ever use grape pH or TA levels to inform a harvest decision, then measurement of these analytes in the vineyard may not be necessary. There may be other points in the production change where pH and TA measurements do inform processing decision, and they can then be applied at that point.

4.9 Conclusions

Chemical analyses can be used throughout the production of alcoholic beverages to ensure and maintain product quality. While many analyses are voluntary, legal requirements exist for many analytes and the appropriate regulatory agencies should be consulted when products are intended to be sold and/or exported. Preparation of a process analytical chart can help to ensure that appropriate analyses are performed throughout the production process. In subsequent chapters we will further discuss the analytical methods that are frequently used for monitoring composition of alcoholic beverages.

References

Bamforth, C.W., 2000. Beer: an ancient yet modern biotechnology. Chem. Educ. 5, 102–112.

Bamforth, C.W., 2016. Brewing Materials and Processes: A Practical Approach to Beer Excellence. Academic Press, London.

Canuti, V., Conversano, M., Li Calzi, M., Heymann, H., Matthews, M.A., Ebeler, S.E., 2009. Headspace solid-phase microextraction-gas chromatography-mass spectrometry for profiling free volatile compounds in Cabernet Sauvignon grapes and wines. J. Chrom. A 1216, 3012–3022.

Cleary, M., Chong, H., Ebisuda, N., Dokoozlian, N., Loscos, N., Pan, B., Santino, D., Sui, Q., Yonker, C., 2015. Objective chemical measures of grape quality. Ebeler, S.E., Sacks, G., Vidal, S., Winterhalter, P. (Eds.), Advances in Wine Research, ACS Symposium Series, vol. 1203, American Chemical Society, Washington, DC, pp. 365–378.

Collins, T.S., Zweigenbaum, J., Ebeler, S.E., 2014. Profiling of nonvolatiles in whiskeys using ultra high pressure liquid chromatography quadrupole time-of-flight mass spectrometry (UHPLC-QTOF MS). Food Chem. 163, 186–196.

Iserentant, D., 1995. Beers: recent technological innovations in brewing. In: Lea, A.G.H., Piggott, J. (Eds.), Fermented Beverage Production. Kluwer, New York, NY, pp. 41–58.

King, E.S., Stoumen, M., Buscema, F., Hjelmeland, A.K., Ebeler, S.E., Heymann, H., Boulton, R.B., 2014. Regional sensory and chemical characteristics of Malbec wines from Mendoza and California. Food Chem. 143, 256–267.

Miracle, R.E., Ebeler, S.E., Bamforth, C.W., 2005. The measurement of sulfur-containing aroma compounds in samples from production-scale brewery operations. J. Am. Soc. Brew. Chem. 63 (3), 129–134.

Muller, R., 1991. The effects of mashing temperature an mash thickness on wort carbohydrate composition. J. Inst. Brew. 97, 85–92.

Muller, R., 2000. A mathematical model of the formation of fermentable sugars from starch hydrolysis during high-temperature mashing. Enzyme Microb. Technol. 27, 337–344.

Nelson, J., Hopfer, H., Gilleland, G., Boulton, R.B., Ebeler, S.E., 2015. Elemental profiling of Malbec wines under controlled winemaking conditions by microwave plasma atomic emission spectroscopy. Am. J. Enol. Vitic. 66 (3), 373–378.

Ough, C.S., Amerine, M.A., 1988. Methods for Analysis of Musts and Wines. Wiley, New York, NY.

Peñas, E., di Lorenzo, C., Uberti, F., Restani, P., 2015. Allergenic proteins in enology: a review of technological applications and safety aspects. Molecules 20 (7), 13144–13164.
Priest, F.G., Stewart, G.G., 2006. Handbook of Brewing, second ed. Taylor & Francis Group, Boca Raton, FL.
Russell, I., Stewart, G., 2003. Whisky: Technology, Production, and Marketing. Elsevier, London.
Scarlata, C.J., Ebeler, S.E., 1999. Headspace solid-phase microextraction for the analysis of dimethyl sulfide in beer. J. Agric. Food Chem. 47 (7), 2505–2508.
Versari, A., Laurie, F., Ricci, A., Laghi, L., Parpinello, G.P., 2014. Progress in authentication, typification and traceability of grapes and wines by chemometric approaches. Food Res. Int. 60, 2–18.

In this chapter

Rapid methods to analyze alcoholic beverages

5.1 Introduction

There are many classical "wet chemistry" methods that are commonly used in an alcoholic beverage analysis laboratory. These methods often measure classes of compounds together (e.g., total solids) rather than individual compounds (e.g., glucose, fructose), although as we will see, this is not always the case. The methods discussed here are usually relatively rapid methods; however, they are often done manually, for example, titrations, although current advances in automation have made many of these methods almost fully automated in many cases. This chapter will not include analytical methods using chromatographic separations as these will be the focus of Chapter 6.

When choosing an analytical method, the analyst should consider the purposes and uses of the analyses as discussed in Chapter 4. As part of this consideration the analyst should consider the needed accuracy, reproducibility, and speed of the analysis. In many cases there may be trade-offs between the accuracy and precision of a given method and its speed, with faster methods often (but not always) being less accurate and/or precise. Organizations, such as the Association of Official Analytical Chemists (AOAC), the American Society of Brewing Chemists (ASBC), the US Alcohol and Tobacco Tax and Trade Bureau (TTB), and International Organisation of Vine and Wine (OIV) provide excellent resources for standard methods for analysis of foods and beverages (ASBC, 2011; AOAC, 2012; OIV, 2011; US TTB, 2016). Methods approved by these organizations undergo thorough testing and evaluation in multiple laboratories before they are approved. This process results in information on the general accuracy and precision of the method, as well as standard sources of errors and interferences in the measurements.

Sensory and Instrumental Evaluation of Alcoholic Beverages. http://dx.doi.org/10.1016/B978-0-12-802727-1.00005-3

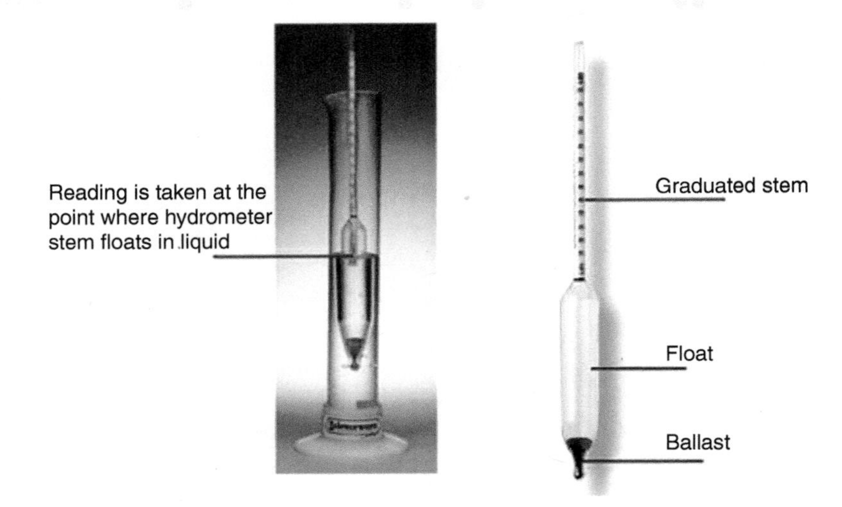

Figure 5.1
A specific gravity hydrometer *http://www.hbinstrument.com/wp-content/uploads/2013/04/p.3-reading-instructions.jpg*

5.2 Soluble solids

The density of a solution is a physical property commonly used to determine the amount of solids dissolved in solution. The main soluble solids in grapes and grape juice are simple monosaccharides, such as glucose and fructose, although organic acids, polysaccharides, proteins/peptides, and salts may also be present (Boulton et al., 1998). In wort, the soluble solids consist mostly of disaccharides and trisaccharides, as well as amylopectins and glucans (Hough et al., 1982; Bamforth, 2000).

Hydrometers provide a rapid method for measuring the density of a solution. A hydrometer is a glass tube with a known weight of mercury or shot in the base and a calibrated scale at the top or neck of the tube (Fig. 5.1). When the hydrometer is placed in a liquid, the hydrometer displaces an equivalent weight of the liquid and this weight can be related to the volume and density of the liquid according to the equation:

$$\text{Density} = \frac{\text{Weight of substance}}{\text{Volume of substance}} \tag{5.1}$$

In practice, the specific gravity of the solution is determined, that is, the relative density of a substance compared to the density of water (1.0000 g/cm^3 at 4°C). As the density of water and all solutions varies with temperature, the actual temperature of the solution should be recorded. Temperature correction charts for hydrometer measurements are used to correct the measured hydrometer value at a given temperature to the actual density at a defined temperature (usually 20°C). As fermentation proceeds, ethanol will influence the density of the solution so that by the time all sugars have been utilized by the yeast, hydrometers calibrated to measure sugar levels cannot be accurately used (e.g., the hydrometer will have a measured specific gravity of less than one).

Soluble solids are usually reported in units of Brix (g sucrose/100 g solution), Balling, degree Plato (°P), degree Baumé, or degree Oechslé (°Oe). Brix, Balling, and degree Plato are similar units of measure with the difference being related to the precision of the specific gravity measurement. The degree Baumé scale is calibrated to give percent salt (NaCl) by mass in a water solution and conversion tables can be used to convert from Brix to Baumé (i.e., 20 Brix is equivalent to 11.1 Baumé; 1.8 Brix = 1 Baumé). In practice, the Baumé reading can provide an estimate of the potential alcohol in the solution (%v/v), for example, for soluble solids with reading of 11.1 Baumé the potential alcohol will be 11.1% if the sugars in the solution are fermented to dryness. Degree Oechslé represents the difference in the specific gravity of the solution relative to the specific gravity of water (1.0000 g/cm^3) multiplied by 1000. For example, the specific gravity of a 20-Brix solution is 1.08287 (at 20°C) corresponding to 82.87 Oechslé.

Recently, density meters have become commercially available that allow rapid measurement of solution density (Brereton et al., 2003; Anton Paar, 2016). These meters have a U-shaped glass tube of known volume and mass that when filled with a solution is oscillated with a piezoelement. The oscillation frequency is inversely proportional to the density of the sample in the tube. The meter is calibrated against air and water and a reference oscillator corrects for temperature variations. The meter is programmed to calculate the density for sugar (and/or alcohol solutions). The meters can be portable and handheld for use in the field, in a processing facility, or on a processing line, or they can be stationary and used in a common laboratory space. The measurements are usually made within a few seconds and the meters are easy to use. The meters should be calibrated periodically to ensure accurate measurements and they usually have a defined range of sample densities, temperatures, and viscosities for which they can be used. Manufacturer instructions should be consulted for operation and maintenance.

When light enters a material it will refract, or bend, and the degree of refraction (i.e., refractive index) is dependent on the material, the temperature of the material, and the wavelength of the light. This physical property can be used to determine the concentration of solutes in solution and refractometers that measure the index of refraction are commonly used to measure the soluble solids content of grapes/grape juice in the field and wort at the beginning of fermentation. Refractometers can be hand-held for use in the field/processing facility or stationary for laboratory use. Digital and automated refractometers are also becoming common for inline process monitoring.

Refractometers are usually calibrated with pure water solutions. If the refractometer does not have automatic temperature compensation, temperature corrections are required. Refractometers require only a few drops of material and as a result are ideal for monitoring harvest maturity of the fruit in the field and for rapid estimates of sugar content of wort. However, once fermentation begins, the refractive index of ethanol in solution significantly contributes to the refraction and so refractometers can no longer be used.

All measures of soluble solids require careful sampling to ensure representative samples are obtained. Particularly with plant materials like grapes and hops, there can be significant variability among plants, clusters of grapes/fruits on a plant or vine, and individual fruit/berries within a cluster. Further variability among fields, growing regions, and years should be considered. The analyst is referred to sources, such as Greenfield and Southgate (2003) for further consideration of general sampling protocols and sources of error.

5.3 Ethanol

The density of pure ethanol is 0.789 g/cm^3 at 20°C and hydrometers calibrated to measure alcohol concentration of solutions can provide a rapid method for determination of the alcoholic strength of alcoholic beverages. Note that the term alcoholic strength is used in the European Union (rather than ethanol concentration) because the standard density-based measurements include other alcohols (e.g., small amounts of methanol) in addition to ethanol in the measurement (European Commission, 1987). For distilled beverages, the sample can often be measured directly because there are few matrix interferences. However, for wine, beer, and liqueur samples, the presence of sugars, acids, proteins, and/or polysaccharides can impact the density of the sample, and these beverages are usually distilled first before using a hydrometer to measure the alcoholic strength.

If the beverage is distilled, a pycnometer, a container with a defined volume, can also be used to assess the weight of the solution at a defined temperature, yielding the density according to Eq. 5.1. From calibration charts, the alcoholic strength of the solution is then derived. Pycnometry is an AOAC approved method for determining ethanol of wines by specific gravity (AOAC, 2012). Alternatively, the refractive index of the distillate can be measured to determine the ethanol concentration (AOAC, 2012).

Ethanol depresses the boiling point of water and this physical property can also be used to measure alcoholic strength of wines and other alcoholic beverages. An ebulliometer is an instrument that heats the sample and measures the temperature at which the ethanol/water mixture begins to boil (a chilled condenser prevents loss of the water/ethanol vapor). Ebulliometers with electric heaters can rapidly heat and regulate the temperature of the sample and digital readouts make it easy to determine the percent alcohol, however, it is difficult to fully automate this analysis.

Ebulliometers usually measure the boiling point of the sample relative to the boiling point of water and frequent calibration is needed (e.g., as atmospheric pressure changes, slight changes in boiling point will occur). The presence of sugar in the sample influences the boiling point and samples with >2% sugar should be diluted prior to analysis (Ough and Amerine, 1988).

Ebulliometers have been used for measuring alcohol in beverages since the 1800s and are still used in many small and medium size facilities. They are relatively easy to use and satisfy the TTB requirement that production facilities have the ability to measure alcohol levels of samples on the premises. However, analysis of ethanol by ebulliometry is not an approved AOAC or OIV method.

5.4 pH

The pH, defined as the −log of the hydrogen ion (H^+) concentration (or activity), of alcoholic beverages can affect perception of sourness and is also important for determining color, chemical reaction rates, and microbial stability. As noted in Chapter 4, pH of grape berries (and most fruits) increases after the onset of ripening and therefore pH is often measured when making harvest decisions.

The pH of a solution is measured with a meter that has an indicating electrode that is sensitive to the hydrogen ion concentration (activity) of the solution and measures the difference in the potential that is created between the outer surface of the indicating pH electrode and the internal surface of a reference electrode. The surface of the pH electrode is glass (i.e., high purity silicon dioxide, SiO_2) and it should be clean and hydrated prior to analysis—that is, the electrode should be stored in aqueous buffer.

Accurate pH measurement requires calibration of the meter on a daily basis—more often if the samples are dirty or highly variable in pH. The meter should be calibrated in the range of pH that will be encountered for a given sample set. For grape samples with pH 3.5–4 at maturity, calibrating between pH 2 and pH 7 is recommended. Most pH meters have automatic temperature compensation but this should be confirmed for the specific meter that is used because temperature can significantly impact hydrogen ion activity. Calibration buffers can be made in house, but it is recommended that these be purchased. The US National Institute of Standards and Technologies (NIST) certifies pH buffers that can be purchased for calibration purposes and this ensures that the buffer is accurate within a defined pH.

Manufacturers of pH meters provide excellent resources on operation, maintenance, and troubleshooting. This information should be made readily available to operators in the laboratory since each manufacturer and/or instrument model number may vary slightly in its use and operation.

5.5 Titratable acidity

Titrations are common volumetric procedures where a known concentration and volume of one solution (the titrant) is reacted with a known volume of another solution in order to determine the concentration of an analyte in the unknown solution. Titrations commonly used for alcoholic beverage analysis rely on acid–base, reduction–oxidation (redox), or complexometric (i.e., titration of a metal ion with a complexing or chelating reagent) reactions.

One of the most common titrations used in alcoholic beverage laboratories is an acid–base titration used to determine titratable acidity. Titratable acidity is defined as the concentration of available protons from all weak acids in the solution (e.g., in the juice, must, wine, beer, and so on) that can be titrated or neutralized by a strong base, such as sodium hydroxide (NaOH). In the United States, titratable acidity of wine is determined by titration of a degassed wine sample to the phenolphthalein endpoint (i.e., pH 8.2) and is reported in tartaric acid equivalents (i.e., 2 moles of hydrogen ions are titrated per mole of tartaric acid). In other countries, the endpoint may be different (e.g., pH 7.0) and may be reported in other equivalent units (e.g., sulfuric acid equivalents). Titratable acidity of beer is typically reported as lactic acid equivalents and is a measure of the extent of lactic acid fermentation, particularly for sour beers. Samples should be degassed to remove dissolved CO_2 (present as carbonic acid, H_2CO_3) from interfering. Autotitrators make it possible to fully automate the titrant additions, equivalence point detection, and all calculations.

Titratable acidity influences the perception of acidity and the physical stability of alcoholic beverages. In wines, the main acid, tartaric acid is present in high concentrations, often near the saturation point, and therefore tartaric acid crystals may precipitate out as the ethanol concentration increases and/or the temperature decreases. In these cases, measurements of titratable acidity are used to determine if acid adjustments are needed to enhance the acidity or if postfermentation procedures will be needed to prevent tartaric acid crystal formation during storage in the bottle. During maturation of fruit, the acid concentration decreases and so both pH and titratable acidity may be monitored to determine optimal harvest maturity.

5.6 Volatile acidity

Volatile acids are defined as the steam-distillable acids present in juice, wine, and beer. The main volatile acid is acetic acid, but lactic, formic, butyric, and propionic acids are included. There are legal limits for volatile acidity and the appropriate regulatory agency should be consulted for the limits for specific alcoholic beverages [e.g., in the United States the Alcohol Tobacco Tax and Trade Bureau regulates volatile acidity levels according to the Code of Federal Regulations (US CFR Title 27, 2016)]. High levels of volatile acidity are typically an indication of spoilage, either from contamination of starting material (e.g., grapes undergoing uncontrolled fermentation between harvest and arrival at the winery), contamination of product with *Acetobacter* species which converts ethanol to acetic acid during fermentation and/or storage, or by uncontrolled malolactic fermentations.

Volatile acidity is determined by steam distilling the sample, collecting the distillate which contains the volatile acids, and titrating the distillate with strong base (sodium hydroxide) to the pH 8.2 (phenolphthalein) endpoint. The results are reported as acetic acid equivalents. When measuring volatile acidity, carbon dioxide (as carbonic acid, H_2CO_3), sulfur dioxide (as sulfurous acid, H_2SO_3), and sorbic acid (added as a preservative to some low alcohol beverages) may interfere in the measurement and if present in high concentrations the measurement should be corrected for these interferences. Careful vacuum degassing can remove dissolved CO_2. SO_2 can be readily corrected for by adding hydrogen peroxide (H_2O_2) prior to distillation. This oxidizes the SO_2 to H_2SO_4, which is not as volatile as SO_2 and is not distilled and collected with the acetic acid distillate. To correct for sorbic acid interference, the sorbic acid concentration is determined separately (e.g., by chromatographic analysis) and the following correction factor is applied: 1 g sorbic acid = 0.535 g acetic acid (Zoecklein et al., 1995). Lactic acid is typically included in measurement of volatile acidity. Only about 2% of the total lactic acid is distilled and collected in the distillate (Zoecklein et al., 1995), however, if lactic acid levels are high, and/or volatile acidity levels are approaching the legal limit, a correction factor may be applied after determining lactic acid levels separately (e.g., by a chromatographic method). Stratification of volatile acids can occur in barrels and tanks so proper sampling and appropriate mixing is necessary to ensure accurate volatile acidity measurements.

Acetic acid can also be measured using an enzyme coupled reaction as described in Section 5.10.3.

5.7 Sulfur dioxide

Sulfur dioxide (SO_2) is added at many stages of alcoholic beverage production as an antioxidant and as an antimicrobial agent. Sulfur dioxide is also formed naturally at low concentrations by microorganisms during fermentation. Any product with greater than 10 mg/L SO_2 must be labeled as containing sulfites to prevent allergic reactions in individuals with SO_2 sensitivity (US TTB). Maximum legal limits for SO_2 also exist and the analyst should consult regulations for the appropriate limits within a given country.

SO_2 exists in multiple forms in alcoholic beverages. Molecular SO_2 (or sulfurous acid, H_2SO_3), bisulfite (HSO_3^-), and sulfite (SO_3^{2-}) are in equilibrium and the amount of each species present is dependent on pH (Eq. 5.2):

$$SO_2(aq;H_2SO_3) \underset{pK_a=1.89}{\rightleftharpoons} HSO_3^- + H^+ \underset{pK_a=7.2}{\rightleftharpoons} SO_3^{2-} + 2H^+ \quad (5.2)$$

Bisulfite also exists in both the "free" form and "bound" to electrophilic species, such as acetaldehyde, pyruvate, reducing sugars, and anthocyanins. The "bound" carbonyl–bisulfite adducts do not provide any antimicrobial or antioxidant activity, however, in some cases the bound adducts are easily disassociated and so can provide a potential source of free SO_2 over the storage time. The reaction of acetaldehyde with SO_2 to form hydroxysulfonate has a strong formation constant (Zoecklein et al., 1995) and if acetaldehyde levels are high, for example, in oxidized wines, the amount of free SO_2 may be low. In red wines, anthocyanins react with SO_2, however, the bound adducts can be readily disrupted by mild heat or hydrolysis over time (Ough and Amerine, 1988).

SO_2 can be measured by a number of methods, with the most common rapid/wet chemistry methods being a redox titration and a distillation followed by acid–base titration. The redox titration (often referred to as the Ripper method) involves titration of the wine with iodine (I_2). During the titration, SO_2 (sulfite) is oxidized to SO_4^{2-} (sulfate) and the endpoint is detected either with a redox electrode or with a starch indicator (at the endpoint of the titration, excess iodine is present and the starch/iodine complex turns blue). Total SO_2 is determined following hydrolysis of the carbonyl–bisulfite adducts with sodium hydroxide, NaOH (pH $>$ 12). To measure free SO_2, the redox titration should be performed quickly, otherwise, the bound adducts will slowly disassociate and the titration endpoint will not be observed. High levels of reducing sugars (e.g., glucose, fructose) in the sample will interfere with the redox titration so this method is not appropriate for wines and alcoholic beverages with high sugar levels.

SO_2 is also frequently measured using the so-called "aeration–oxidation" procedure. This is a modification of the AOAC approved Monier-Williams method for total SO_2 analysis (AOAC, 2012). SO_2 is volatilized/distilled from an acidified sample by purging with air or pulling a vacuum. The purged SO_2 is collected in a flask containing hydrogen peroxide (H_2O_2). The hydrogen peroxide oxidizes SO_2 to sulfuric acid (H_2SO_4) which is titrated with sodium hydroxide. A chilled condenser helps to prevent collection of acetic acid in the collection flask. To measure free SO_2, the distillation conditions (time, temperature, vacuum pressure, or purge flow rate) are controlled in order to minimize disassociation of the carbonyl–bisulfite adducts. Total SO_2 is measured by heating the sample during the distillation step to hydrolyze all bound adducts.

Both the Ripper and aeration–oxidation methods provide only an approximation of the free SO_2 concentration since the carbonyl–bisulfite adducts are readily disrupted during the analysis. However, when done carefully, both methods can provide an accurate measurement of total SO_2 in alcoholic beverages.

5.8 Nitrogen, NH_3/NH_4^+, and α-amino acids

Yeast requires nitrogen as a nutrient during fermentation. Ammonia (NH_3) is a readily assimilable form of nitrogen and α-amino acids can also be utilized. There is no direct correlation between NH_3 concentration and α-amino acid concentration in grapes (Butzke, 1998) and in most cases, both forms of nitrogen should be measured to ensure sufficient nitrogen to prevent stuck fermentations.

Ammonia can be measured using an ammonia-specific electrode attached to a pH meter or voltmeter. The indicating electrode is a nylon membrane that allows NH_3 (gas) to pass through. Once in the interior of the electrode, the pH is acidic and NH_3 is converted to ammonium, NH_4^+. The change in internal pH is sensed with an internal pH electrode; by using standard solutions of ammonium chloride (NH_4Cl), the meter response can be calibrated and used to determine the ammonia concentration in the original sample (McWilliam and Ough, 1974).

The formol titration measures both ammonia and α-amino acids. Formaldehyde is added to the sample and reacts with the amino acids and ammonia, releasing a hydrogen ion (H^+) and causing the pH of the solution to drop. The released protons are titrated with NaOH to the phenolphthalein (pH 8.2) endpoint and the amount of titrant is proportional to the total nitrogen from ammonia and amino acids in the sample. Not all amino acids react with formaldehyde to the same extent due to their different basicities. Therefore, in a complex mixture of nitrogen and amino acids, the formol titration can only yield an approximation of the total nitrogen content (Zoecklein et al., 1995).

α-Amino acids can also be measured spectroscopically and will be discussed in Section 5.10.2.

5.9 Water hardness and metals/salts

Complexometric titrations are often used for determination of water hardness, which is important for brewing, distillation, and general cleaning operations in all processing facilities (Eumann and Schaeberle, 2016). To determine water hardness, ethylenediaminetetraacetic acid (EDTA) of known concentration is used as a titrant to complex with metal ions in solution, predominantly Ca^{2+} and Mg^{2+}. Metal ion indicators, such as Eriochrome Black T and Calmagite are often used to detect the endpoint and the results are reported as $CaCO_3$ equivalents. The titration is usually performed at pH 10 since at higher pH, metal hydroxides will precipitate [e.g., $Mg(OH)_2$] and cannot be complexed with EDTA (Christian, 2004).

Similar to pH electrodes, specific ion electrodes are also available for measurement of cations (e.g., K^+, Na^+, and so on) and anions (e.g., Cl^- and so on) in a sample. As with pH meters, these electrodes require calibration and manufacturer directions to provide the best resource for operation and maintenance.

Atomic spectroscopy methods (e.g., atomic absorption spectroscopy, inductively coupled plasma mass spectrometry, and so on) for measuring elements will be discussed in Chapter 6.

5.10 Molecular spectroscopy methods

In this final section we will discuss molecular spectroscopy methods for analysis of alcoholic beverages. These include ultraviolet (UV)-visible spectroscopy–based methods and infrared-based methods.

Spectroscopic methods measure interactions of molecules or atoms with electromagnetic radiation. With molecular spectroscopy methods, molecules absorb energy and are raised to a higher energy state. The amount of energy absorbed will determine the type of molecular transition. When energy in the far infrared and/or microwave region (wavelengths of ~100 μm) is absorbed by molecules it increases the rotational energy of the molecule and the molecules begin to rotate about various axes. For energy in the infrared region (wavelengths of ~1000 nm), vibrational transitions occur and groups of atoms in a molecule vibrate relative to each other. These transitions are the rocking and scissoring types of vibrations measured in infrared spectra. With high energy wavelengths in the UV and visible regions (200–800 nm), electrons in bonding and outer valence molecular orbitals can be excited and these electronic transitions are commonly used for UV-visible spectroscopic measurements of molecules in solutions.

Only molecules with a chromophore will absorb in the UV-visible region; nonbonding electrons and pi-bonding electrons (i.e., double bonds) provide the strongest UV-visible absorbing chromophores. Highly conjugated systems are also easily excited and readily absorbed in the UV-visible region. If an analyte does not have a chromophore, it can be reacted with another compound to form a product species that absorbs in the UV-visible region. Enzyme-coupled reactions are also commonly used. Examples of each of these analyses are described in the following sections.

UV-visible molecular spectroscopy relies on the Beer–Lambert law to relate the absorption of energy (A) at a specific wavelength (λ) to the concentration (c) of the analyte of interest:

$$A = a \times b \times c \tag{5.3}$$

where A is the absorption of light at a specific wavelength; a is the molar absorptivity or extinction coefficient (ε) of the analyte (IUPAC currently recommends the term "molar attenuation coefficient" to take into account effects of scattering and luminescence, IUPAC (1997)); b is the path length or distance the light energy travels through the sample; and c is the concentration of the sample. If the absorptivity/extinction coefficient is known, analyte concentration can be determined directly from the absorbance of the unknown sample using the Beer–Lambert law. Alternatively, calibration curves can be created where the absorbance of a set of standards is plotted as a function of the concentration of the

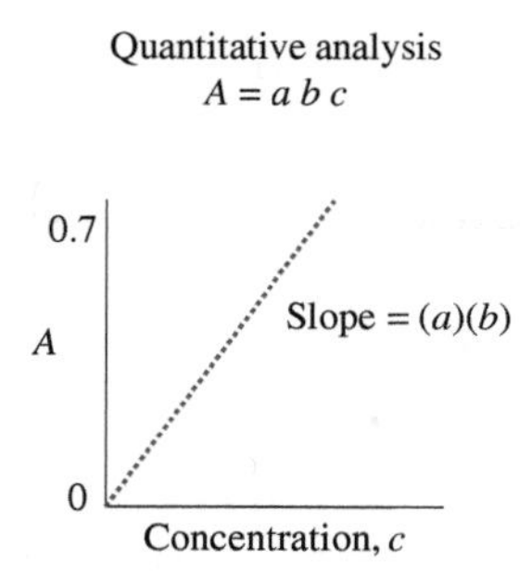

Figure 5.2
A calibration curve for a UV-visible spectrophotometric analysis
The absorbance of a standard containing the analyte of interest is determined at a defined wavelength and the measured absorbance (A) is plotted as a function of analyte concentration (c). The slope of the calculated regression line *(dotted line)* is equivalent to the absorbance (a) or molar extinction coefficient (ε) depending on the concentration units. The path length (b) is typically 1 cm and is the same for all samples and standards. Analyte concentrations in an unknown sample can be determined by measuring the absorbance (y-axis) and solving the regression equation ($y = mx + b$) to determine the concentration, c (x-axis).

standards (Fig. 5.2). The path length should remain constant for all samples in the calibration and test sets. From the calibration curve a regression equation is calculated. The slope is proportional to the absorptivity (or molar extinction coefficient, ε) multiplied times the path length.

When making measurements, the sample container (also called the sample cell) should not absorb light at the wavelength of interest; at low UV wavelengths this will require use of quartz cells which are UV transparent. The samples should be filtered to remove particulates and sample dilution may be required to ensure that the measurements are within the optimal range of the instrument (~0.2–0.8 absorbance units; analysts should refer to instrument manuals for operating instructions).

5.10.1 Total phenolics, anthocyanins, and color

Phenolics in alcoholic beverages contribute to bitter taste and astringent mouthfeel (i.e., a puckering, drying sensation in the mouth). Anthocyanins are a class of phenolics that contribute to grape and wine color. Phenolics can combine to form a number of oligomeric and polymeric structures—the polymeric compounds are frequently called "tannins." Phenolics also have antioxidant activity and red wines, which typically have 1500–2500 mg/L total phenolics, can age for longer times without significant oxidative degradation compared to white wines (Boulton et al., 1998).

Phenolics have an aromatic ring that will readily absorb UV light at 280 nm, therefore absorption at 280 nm is often used as a measure of total phenolic content (Somers and Evans, 1977; Harbertson and Spayd, 2006) (Fig. 5.3). Anthocyanins, such as malvidin-3-glucoside, which give wine its red color have a strong absorption at 520 nm, while other phenolic classes will absorb at other wavelengths, for example, 365 nm for flavonols like quercetin (Fig. 5.3). Direct measurement at specific wavelengths does not provide an accurate determination of phenol concentration, however, because the different classes of phenols and compounds within a class have different molar extinction coefficients at the different wavelengths making it difficult to calibrate the measured absorbance with known standards. By standardizing to constant pH, by comparing absorbance with and without added SO_2 (SO_2 bleaches monomeric phenols, but not polymeric phenols), by measuring absorbance at different sample dilutions, and by estimating the extinction coefficients (ε), spectral measurements can be used to determine concentrations of monomeric and polymeric anthocyanins, total phenols, and copigmentation in alcoholic beverages (Boulton, 2001; Levengood and Boulton, 2004).

Total phenolics can also be measured by reaction with a reagent to form complexes that absorb in the visible wavelength region. FeCl (ferric chloride) will react with phenols to form a colored complex that absorbs at 510 nm. By adding a protein (e.g., bovine serum albumin), polymeric phenols bind to the protein and precipitate from solution. The precipitate is isolated by centrifugation and resuspended in buffer and after reacting with FeCl, tannin (polymeric phenol) concentration can be determined (Harbertson et al., 2003). (+)-Catechin is used for calibration and responses are reported in (+)-catechin equivalents. This method is sometimes called the Harbertson–Adams assay.

The Folin–Ciocalteu method is based on the chemical reduction of the Folin reagent (a phosphomolybdo-tungsten hetero acid) by phenols in solution (Singleton and Rossi, 1965). The resulting reduced Folin reagent absorbs light at 765 nm. Standard solutions of gallic acid are used to calibrate the spectrophotometer response across a range of concentrations. Using the regression equation from the calibration standards, the total phenol concentration of an unknown sample can be determined. Or alternatively, the extinction coefficient of the reagent can be used to calculate the unknown concentrations based on the absorbance of the sample at 765 nm. The Folin–Ciocalteu method is widely used for food and beverage analysis of total phenols, but other reducing agents (e.g., sugars) can interfere with the analysis and the method is not appropriate for sweet wines or samples with high SO_2 levels relative to the total phenol concentration. See also Singleton and Rossi (1965) for a thorough review of factors influencing measurement of total phenols by the Folin–Ciocalteu method.

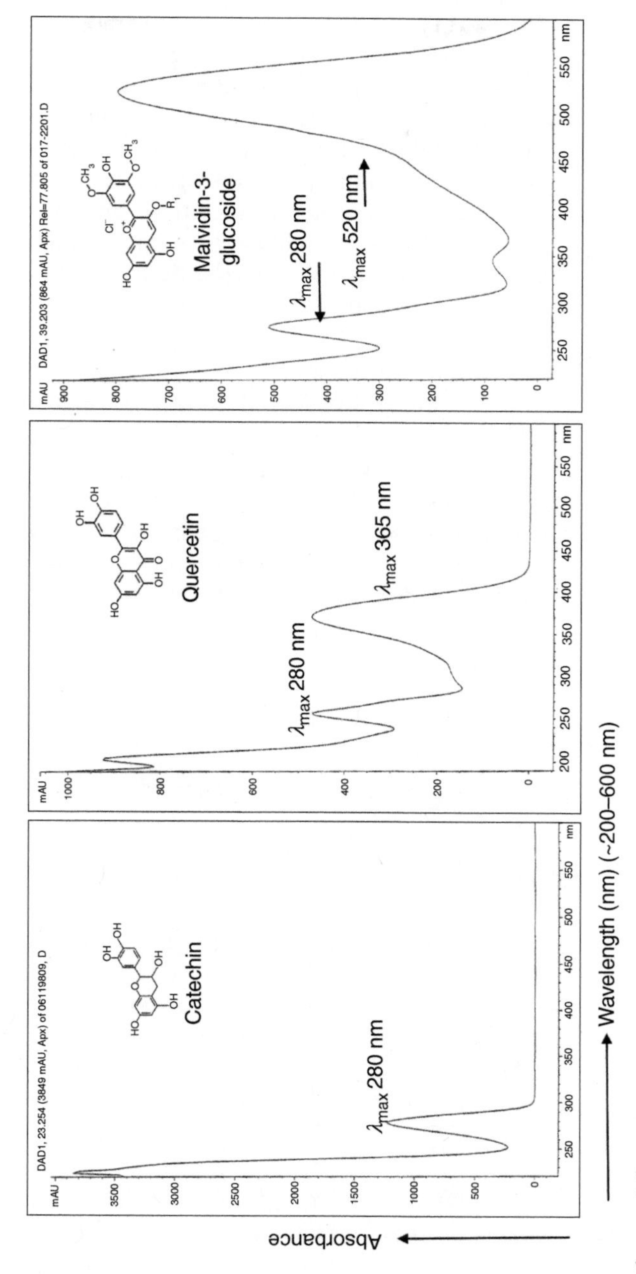

Figure 5.3
UV-visible absorption spectra for selected phenolics

The color of wort, malt, and beer are measured spectroscopically at 430 nm (ASBC, 2011). The color is reported as Standard Reference Method values which are obtained by multiplying the absorbance of the sample at 430 nm by the sample dilution and a standard correction factor of 12.7. The correction factor represents the constant value used to relate the absorbance measurement to standard color tiles used in early published methods (called Lovibond). The Standard Reference Method colors range from 2 for a light pale ale to greater than 40 for dark stouts. The color in wort, malt, and beer derives from Maillard reactions (i.e., browning reactions between sugars and amino acids) in the malt that occur during toasting. The color of aged distilled beverages, for example, whiskeys, can also be measured in the same manner.

5.10.2 α-Amino acids

α-Amino acids absorb energy in the very low UV region (<190 nm), however numerous other analytes interfere at these wavelengths. Therefore, the amino acids can be quantitatively reacted with another reagent to yield a product with a strong UV-active chromophore which can be readily measured and quantified. A common reactant is *o*-phthaldialdehyde (OPA) which, in the presence of *N*-acetyl cysteine (NAC), reacts with the primary amino group of amino acids to form an isoindole chromagen which absorbs at 335 nm (Dukes and Butzke, 1998) (Eq. 5.4).

$$\text{OPA} + R_1SH \xrightarrow[\alpha\text{-Amino acid}]{R_2NH_2} \text{Isoindole chromagen (S-}R_1\text{, N-}R_2\text{) A335} \tag{5.4}$$

OPA NAC Isoindole chromagen A335

The amino acid proline does not readily react and is not measured in this assay. Since yeast do not readily use proline as a nitrogen source, this spectroscopic method can give a good estimate of the α-amino content of juice which provides a nitrogen source for the yeast. Ammonia/ammonium (NH_4^+) has only a very weak reaction with OPA/NAC and so for determination of total available nitrogen, ammonia/ammonium concentration should be determined separately (Dukes and Butzke, 1998). Standard calibration curves are prepared with isoleucine and results are presented as α-amino acid concentration in isoleucine equivalents. Hydroxycinnamate phenols absorb in the wavelength region and careful correction for background matrix interferences is required.

5.10.3 Enzymatic-coupled reactions

If an analyte does not have a chromophore, an enzyme can be used to catalyze a chemical reaction of the analyte in which a direct product or cofactor in the reaction has a UV-visible active chromophore. If the reactions are complete and stoichiometric, the concentration of the UV-visible active compounds that is formed can be directly related to the initial analyte concentration. Common reactions used for quantitative analysis purposes involve dehydrogenase enzymes and interconversion of $NADP^+/NADPH$ or $NAD^+/NADH$ (Eq. 5.5):

$$A + B \xrightarrow[NADP^+ \rightarrow NADPH]{\text{Dehydrogenase}} C + D \quad (5.5)$$

NADPH (NADH) absorbs light at 340 nm, while $NADP^+$ (NAD^+) does not. Therefore, by measuring an increase or decrease in absorbance at 340 nm, and using the molar extinction coefficient of NADP/H (6220 M^{-1} cm^{-1}) the molar concentration of the original analyte can be obtained. As shown in the example Eq. 5.5, 1 mole of NADPH is formed by the dehydrogenase enzyme in the complete reaction of 1 mole of Analyte A to products C and D. From the stoichiometry of this reaction, the concentration of Analyte A in the original sample can be calculated.

A number of analytes can be measured with enzyme-linked assays using dehydrogenase enzymes, including glucose, fructose, sucrose, ethanol, acetic acid, malic acid, lactic acid, and ammonia. Commercial manufacturers provide kits with all the reagents needed for a specific analysis (Anon, 2016). Enzyme-coupled reactions are usually relatively rapid and easy to perform and can be automated with robotic analyzers. The enzyme-coupled methods can be highly sensitive to low analyte concentrations and the enzymes are usually specific to the analytes of interest, so interferences may be minimal. However, the assays require careful control of pH, temperature, and ionic strength conditions for optimal enzyme activity. Due to the low concentrations that can be measured, small volumes and extensive dilutions may be needed, which can introduce dilution errors. The specificity of enzymes may vary and in some cases, multiple analytes may be measured (e.g., alcohol dehydrogenase converts ethanol to acetaldehyde, the enzyme reacts with a range of low molecular weight alcohols, and so the analysis is not necessarily specific for ethanol). Phenolics can absorb in the same region as NADP/H and so background subtractions may be needed to correct for absorption due to phenolics in the sample matrix. Interfering phenolics can also be removed from the sample by treating with charcoal or polyvinylpolypyrrolidone (PVPP). As with all analyses, careful calibration and standardization is needed for optimal results.

5.10.4 Infrared spectroscopy and sensor-based methods

There is increasing interest in analytical methods that can measure multiple analytes rapidly, simultaneously, and nondestructively. Many of these methods rely on a range of types of chemical sensors which respond differentially to individual analytes or classes of analytes. The analysis is based on patterns of sensor responses to the complex sample matrix and from the responses a quantitative analysis of specific analytes can be obtained. Infrared analyzers are increasingly used for analyses such as this. Cozzolino et al. (2011) and Bauer et al. (2008) provide an excellent overview of the application of near infrared (~800–2,500 nm) and midinfrared (~2,500–50,000 nm) spectroscopy for quantitative analysis of a range of analytes in grape juice and wines. The methods rely on analysis of large numbers of samples and chemometric data analysis (e.g., principal components analysis; partial least squares regression, and so on) to relate the infrared absorption spectrum of the sample to analyte concentrations determined by a standard, primary method. As the IR methods are based on comparison to a primary analytical method, they are often considered "indirect" or "secondary" methods. These indirect spectroscopic techniques have been used to quantify polyphenols, anthocyanins, tannins, polysaccharides, sugars, alcohols, organic acids, pH, and SO_2 in a variety of alcoholic beverages (Kupina and Shrikhande, 2003; Lachenmeier, 2007; Grassi et al., 2014). Commercial infrared spectroscopic instruments are available.

Other types of sensor-based methods rely on UV-visible spectra to measure color and phenolic concentrations (Skogerson et al., 2007; Shrake et al., 2014); metal oxide and polymer-based sensors to measure volatiles (Lozano et al., 2015; Han et al., 2016); and peptide-based sensors to measure organic acids, polyphenols, and tannins (Ghanem et al., 2015).

All indirect spectroscopic methods require careful instrument calibration and analysis of large numbers of samples to model the full range of analyte concentrations and matrix conditions. Errors can occur if a sample is analyzed that falls outside the range of samples used to develop the calibration models. For example, if a sensor-based analysis is used to measure the ethanol concentration of a red wine, the calibration models for determining ethanol concentration in an unknown sample should include analysis of a range of red wine samples with varying alcohol concentrations and phenolic/color concentrations. Analysis of a white wine using calibrations for red wines may result in inaccurate measurements.

5.11 Ensuring validation of chemical methods and developing standard operating procedures

Fundamental principles of analytical chemistry require that the analytical methods used in an alcoholic beverage testing laboratory be appropriate for the intended application (CITAC and Eurachem, 2002; Harris, 2003; Christian, 2004). General steps to ensure that an analytical laboratory is producing data that are fit to purpose and are of high quality include the following: (1) develop and use an analytical process chart; (2) use validated methods; (3) have well-defined quality control and quality assurance processes and procedures; (4) document problems and failures in the laboratory and with analytical methods; and (5) have written standard operating procedures (SOPs). With the exception of the analytical process chart which was discussed in Chapter 4, these steps are further discussed as follows.

Method validation is performed to demonstrate that a given procedure meets the expected specifications for the intended purpose (Harris, 2003). The analytical process chart that is developed can be used to help guide the validation process. When doing a method validation, it is important to remember that every laboratory is unique and even previously published protocols should be validated in your own laboratory and with your own analysts performing the analyses. This includes ensuring that the method has the level of accuracy and precision (reproducibility) needed and that it has the specificity and selectivity needed. The method validation can include evaluation of matrix effects and determination of method limits (e.g., the linear response range of an instrument, the lowest limits of detection and quantification, the influence of matrix interferences, and so forth). The validation should include documentation and written protocols describing the validation process.

A complete Quality Control/Quality Assurance program helps to ensure the reliability of specific samples or batches of samples and provides a general description of the quality procedures and requirements for a laboratory (Sargent and Hammond, 1997). This includes running control samples at the same time as the analytical samples. Control samples should have a well-defined composition and are treated identically to the test or unknown samples. The responses of the control samples (e.g., measured analyte concentration) are monitored periodically throughout a run and over a series of days. If a sample falls outside of the specifications expected for the control sample, the analyst should stop the analysis, troubleshoot, identify, and correct potential problems, and then reanalyze the control samples. The troubleshooting and problem-solving steps are repeated until the response for the control sample again falls within the expected control limits.

General laboratory quality assurance principles include providing expectations for staff training and ongoing education; specifications for all chemicals and reagents used; specifications for equipment and documents for calibration and maintenance procedures; documented methods and quality control procedures; and documented data reporting procedures. Accreditation and proficiency testing may be components of a full quality assurance program but in and of themselves they are not enough to ensure that high quality data are produced in an analytical laboratory.

Documentation of method failures and problems is critical for building a knowledge base about a given method. Since failures and problems are rarely unique, the database can help to troubleshoot problems more efficiently. Documentation may also allow issues to be identified earlier, which may make it easier to solve the problem, and/or to prevent long and costly downtime due to significant and/or catastrophic instrument or method failures.

Finally, developing detailed, written SOPs for every laboratory activity can help to ensure consistent and high quality performance of procedures and processes within a laboratory. This includes SOPs for sample chain of custody and handling, instrument maintenance, details of the analytical methods and procedures, specifications for recording and archiving data and reports, and safety procedures and considerations. The SOPs will be specific to the organization and laboratory where they are developed and used. They should be written as concisely as possible in a step-by-step format and made readily available to all who will need and use them.

5.12 Conclusions

The rapid methods described in this chapter provide general compositional information for alcoholic beverages. The methods are based on fundamental physical and/or chemical properties of the analytes and most methods discussed in this chapter measure only one compound or a class of compounds at a time. Many of the methods described here can be automated or semiautomated leading to improvements in precision and speed of analysis. Rapid, sensor-based methods allow for multiple analytes to be measured simultaneously in only a few seconds or minutes, however, these methods are indirect methods and require extensive validation with large data sets using samples with similar matrices to those of the test samples. For all methods, standard laboratory approaches are needed to ensure that the data produced by an analytical laboratory are appropriate for the intended purpose. In Chapter 6 we will expand the discussion of chemical analyses to include instrumental methods for volatiles, nonvolatiles, and elements, with a focus on targeted and nontargeted chromatographic approaches.

References

Anon, 2016. Enzymatic Analysis Test Kits. Available from: http://www.r-biopharm.com/products/food-feed-analysis/constituents/enzymatic-analysis

Anton Paar. Density Meters. Available from: http://www.anton-paar.com/us-en/products/group/density-meter/

Association of Official Analytical Chemists (AOAC), 2012. Official Methods of Analysis of AOAC International, nineteenth ed. AOAC International, Gaithersburg, MD.

American Society of Brewing Chemists (ASBC), 2011. Methods of Analysis, fourteenth ed. American Society of Brewing Chemists, St. Paul, MN, USA.

Bamforth, C.W., 2000. Beer: an ancient yet modern biotechnology. Chem. Educ. 5, 102–112.

Bauer, R., Nieuwoudt, H., Bauer, F., Kossmann, J., Koch, K., Esbensen, K.H., 2008. FTIR spectroscopy for grape and wine analysis. Anal. Chem. 80, 1371–1379.

Boulton, R., 2001. The copigmentation of anthocyanins and its role in the color of red wine: a critical review. Am. J. Enol. Vitic. 52, 67–87.

Boulton, R.B., Singleton, V.L., Bisson, L.F., Kunkee, R.E., 1998. Principles and Practices of Winemaking. Springer, New York, NY.

Brereton, P., Hasnip, S., Bertrand, A., Wittkowski, R., Guillou, C., 2003. Analytical methods for the determination of spirit drinks. Trends Anal. Chem. 22 (1), 19–25.

Butzke, C.E., 1998. Survey of yeast assimilable nitrogen status in musts from California, Oregon, and Washington. Am. J. Enol. Vitic. 49 (2), 220–224.

Christian, G.D., 2004. Analytical Chemistry. John Wiley, New York, NY.

CITAC, Eurachem, 2002. Guide to Quality in Analytical Chemistry. CITAC, Eurachem, Available from: https://www.eurachem.org/index.php/publications/guides/qa And a pdf of the Guide to Quality in Analytical Chemistry: https://www.eurachem.org/images/stories/Guides/pdf/CITAC_EURACHEM_GUIDE.pdf

Cozzolino, D., Cynkar, W., Shah, N., Smith, P., 2011. Technical solutions for analysis of grape juice, must and wine: the role of infrared spectroscopy and chemometrics. Anal. Bioanal. Chem. 401 (5), 1475–1484.

Dukes, B.C., Butzke, C.E., 1998. Rapid determination of primary amino acids in grape juice using an *o*-phthaldialdehyde/*N*-acetyl-L-cysteine spectrophotometric assay. Am. J. Enol. Vitic. 49 (2), 125–134.

Eumann, M., Schaeberle, C., 2016. Water. In: Bamforth, C.W. (Ed.), Brewing Materials and Processes: A Practical Approach to Beer Excellence. Academic Press, London, pp. 97–111.

European Commission, 1987. Commission Directive 87/250/EEC on the indication of alcoholic strength by volume in the labelling of alcoholic beverages for sale to the ultimate consumer. Off. J. Europ. Comm. 113, 57–58.

Ghanem, E., Hopfer, H., Navarro, A., Ritzer, M.S., Mahmood, L., Fredell, M., Cubley, A., Bolen, J., Fattah, R., Teasdale, K., Lieu, L., Chua, T., Marini, F., Heymann, H., Anslyn, E.V., 2015. Predicting the composition of red wine blends using an array of multicomponent peptide-based sensors. Molecules 20 (5), 9170–9182.

Grassi, S., Lyndgaard, C.B., Foschino, R., Amigo, J.M., 2014. Beer fermentation: monitoring of process parameters by FT-NIR and multivariate data analysis. Food Chem. 155, 279–286.

Greenfield, H., Southgate, D.A.T., 2003. Sampling. Food Composition Data. Food and Agriculture Organizations of the United Nations, Rome, (Chapter 5).

Han, J., Bender, M., Seehafer, K., Bunz, U.H.F., 2016. Identification of white wines by using two oppositely charged poly(*p*-phenyleneethynylene)s individually and in complex. Angew. Chem. Int. Ed. 55, 1–5.

Harbertson, J.F., Picciotto, E.A., Adams, D.O., 2003. Measurement of polymeric pigments in grape berry extracts and wines using a protein precipitation assay combined with bisulfite bleaching. Am. J. Enol. Vitic. 54, 301–306.

Harbertson, J.F., Spayd, S., 2006. Measuring phenolics in the winery. Am. J. Enol. Vitic. 57, 280–288.

Harris, D.C., 2003. Quantitative Chemical Analysis. W.H. Freeman, New York, NY.

Hough, J.S., Briggs, D.E., Stevens, R., Young, T.W., 1982. Malting and Brewing Science, Hopped Wort and Beer, vol. II. Chapman and Hall, New York.

International Organisation of Vine and Wine (OIV), 2011. Compendium of International Methods of Wine and Must Analysis. OIV, Paris.

IUPAC, 1997. Compendium of Chemical Terminology, second ed. (the "Gold Book"). Compiled by A. D. McNaught and A. Wilkinson. Blackwell Scientific Publications, Oxford. XML online corrected version: http://goldbook.iupac.org (2006–) created by M. Nic, J. Jirat, B. Kosata; updates compiled by A. Jenkins. Available from: http://goldbook.iupac.org/A00516.html

Kupina, S.A., Shrikhande, A., 2003. Evaluation of a Fourier transform infrared instrument for rapid quality-control wine analyses. Am. J. Enol. Vitic. 54, 131–134.

Lachenmeier, D.W., 2007. Rapid quality control of spirit drinks and beer using multivariate data analysis of Fourier transform infrared spectra. Food Chem. 101, 825–832.

Levengood, J., Boulton, R.B., 2004. The variation in the color due to copigmentation in young Cabernet Sauvignon wines. In: Waterhouse, A.L., Kennedy, J.A. (Eds.), Red Wine Color: Exploring the Mysteries, ACS Symposium Series, vol. 886. American Chemical Society, Washington, DC, pp. 35–62.

Lozano, J., Santos, J.P., Suárez, J.I., Cabellos, M., Arroyo, T., Horrillo, C., 2015. Automatic sensor system for the continuous analysis of the evolution of wine. Am. J. Enol. Vitic. 66 (2), 148–155.

McWilliam, D.J., Ough, C.S., 1974. Measurement of ammonia in musts and wines using a selective electrode. Am. J. Enol. Vitic. 25 (2), 67–72.

Ough, C.S., Amerine, M.A., 1988. Methods for Analysis of Musts and Wines. Wiley, New York, NY.

Sargent, M., Hammond, J., 1997. Designing analytical methodology for the generation of data that are "Fit for Purpose". Spectroscopy 12 (7), 46–49.

Shrake, N.L., Amirtharajah, R., Brenneman, C., Boulton, R., Knoesen, A., 2014. In-line measurement of color and total phenolics during red wine fermentations using a light-emitting diode sensor. Am. J. Enol. Vitic. 65 (4), 463–470.

Singleton, V.L., Rossi, Jr., J.A., 1965. Colorimetry of total phenolics with phosphomolybdic-phosphotungstic acid reagents. Am. J. Enol. Vitic. 16, 144–158.

Skogerson, K., Downey, M., Mazza, M., Boulton, R., 2007. Rapid determination of phenolic components in red wines from UV-visible spectra and the method of partial least squares. Am. J. Enol. Vitic. 58, 318–325.

Somers, T.C., Evans, M.E., 1977. Spectral evaluation of young red wines: anthocyanin equilibria, total phenols, free and molecular SO_2 and chemical age. J. Sci. Food Agric. 28, 279–287.

US Code of Federal Regulations Title 27, Alcohol, Tobacco Products and Firearms, 2016. Available from: http://www.ecfr.gov/cgi-bin/text-idx?SID=f5eb0916bf534618be5938f55de1c709&mc=true&tpl=/ecfrbrowse/Title27/27tab_02.tpl

US Alcohol, Tobacco Tax and Trade Bureau. Available from: https://www.ttb.gov

Zoecklein, B., Fugelsang, K.C., Gump, B.H., Nury, F.S., 1995. Wine Analysis and Production. Chapman and Hall, New York, NY.

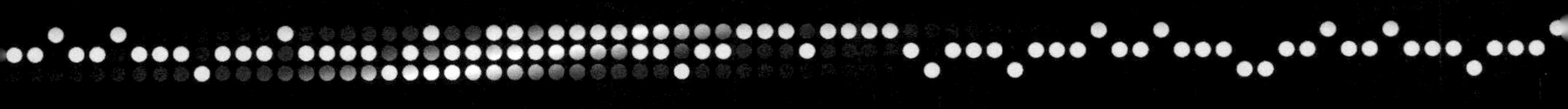

In this chapter

Instrumental analyses for alcoholic beverages

6.1 Introduction

In Chapter 5 we provided an overview of common rapid, wet chemistry methods for alcoholic beverage analysis. In the current chapter we focus on instrumental approaches that can be used to provide compositional analysis of volatile, nonvolatile, and elemental components. We focus on chromatographic approaches and mass spectrometry (MS) but also provide a brief discussion of applications of capillary electrophoresis (CE) and nuclear magnetic resonance (NMR) analyses. For each instrumental method we provide a brief introduction and overview as well as selected applications to alcoholic beverages. A variety of books can provide a more detailed discussion of the theory, practice, and applications of these methods (Harris, 2003; Lea and Piggott, 2003; Skoog et al., 2007; Poole, 2012; Russell and Stewart, 2014; Bamforth, 2016).

6.2 Chromatography

Chromatography is defined as the separation of a mixture of two or more analytes into individual components based on the differential distribution of the components between a stationary phase and a mobile phase. The types of chromatography can be distinguished based on the nature of the mobile and stationary phases, on the nature of the analytes separated, and/or on the equilibration processes involved in the separation, including adsorption, partitioning, ion exchange, and size exclusion mechanisms (Fig. 6.1). Further explanation of these mechanisms will be discussed in the sections below with example applications provided. Chromatographic approaches can be used to identify unknown compounds and to quantify individual components in a sample. Sample preparation is often a key consideration for chromatographic analyses and will be discussed in more detail in Section 6.5.

Chromatographic instruments include an inlet which is used to introduce the sample to the analytical column where the separation occurs, as well as a detector

Sensory and Instrumental Evaluation of Alcoholic Beverages. http://dx.doi.org/10.1016/B978-0-12-802727-1.00006-5

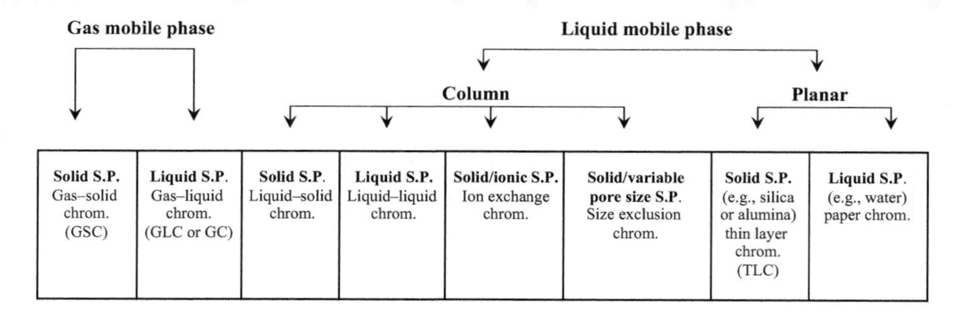

Figure 6.1 Types of chromatography based on physical properties of the mobile and stationary phase Abbreviation S.P. refers to stationary phase.

which converts the chemical signal into an electronic signal as analytes elute from the column. The signal response is usually recorded digitally with a computer. With gas chromatography (GC), the analytical column is contained in an oven which allows the sample to be heated to volatilize the samples. Gas or liquid flow meters are required to control the flow of the mobile phase. The types of detectors used will depend on the chromatographic method and analytes measured, but increasingly mass spectrometer detectors are combined with chromatographic separations and will be discussed in Section 6.3. Most chromatographic instruments are equipped with an autosampler to ensure automatic, precise, and accurate injections for large number of samples as well as a computer which is used to control the instrument and record the detector responses.

6.2.1 Gas chromatography

GC, developed by Cremer and James and Martin in the 1940s and 1950s, is used for analysis of volatile compounds (Ettre, 2008; Jennings and Poole, 2012). Typically, this includes analytes with molecular weight of less than ~600 Da and boiling points of less than ~500°C (the compounds must be thermally stable at these elevated temperatures) (Dettmer-Wilde and Engewald, 2014). Aroma compounds in alcoholic beverages are volatile and most commonly analyzed by GC, therefore GC is often used to relate sensory aroma to volatile aroma composition. This will be discussed further in Section 6.9. In addition, GC analysis of ethanol is a standard AOAC approved method (AOAC, 2012).

Some polar, nonvolatile components can also be analyzed by GC but require derivatization, which results in decreasing the boiling point and/or polarity of the analyte. We use the terms polarity, nonpolar, and polar, here to refer to the ability of a molecule to form dipoles, induced dipoles, and or partial charges. When bonding electrons are equally shared between two atoms, the bond is a nonpolar covalent bond. When the electrons are not equally shared, there can be a partial transfer of electron density from one atom to another, this can lead to formation of dipoles and depending on the electronegativity of the atoms involved a partial charge on the molecule. When a molecule has bond dipoles that do not cancel out, molecular dipoles are created leading to increased polarity of the molecule.

Fatty acid analysis is a good example of a group of compounds that can be analyzed by GC—but their highly polar nature makes their separation difficult and they typically have poor chromatographic behavior with poor peak shapes. As a result, fatty acids are usually derivatized to form fatty acid methyl esters (Eq. 6.1) which are more nonpolar relative to the free acids and can be readily analyzed by GC.

$$\underset{\text{Fatty acid}}{R_1C(=O)OH} + \underset{\text{Methanol}}{CH_3OH} \longrightarrow \underset{\text{Fatty acid methyl ester}}{R_1C(=O)OCH_3} + \underset{\text{Water}}{H_2O} \tag{6.1}$$

Separation mechanism for GC are typically based on adsorption of analytes to the surface of solid particles in the column [gas–solid chromatography (GSC)] and/or partitioning of the analytes between the gas phase and a liquid phase coated on the column surface (or coated on the surface of a support material such as diatomaceous earth) [gas–liquid chromatography (GLC)]. The mobile phase is an inert gas, usually hydrogen, helium, or nitrogen, and the flow is controlled with pressure and/or flow controllers. By heating the column in the GC oven, analytes volatilize at temperatures and pressures corresponding to their vapor pressures. As a result, analytes are separated based on their vapor pressures as well as their chemical/physical interactions with the stationary phase.

GSC with solid and porous solid supports has been used for separation of light gases [e.g., separation of hydrogen sulfide (H_2S), dimethyl sulfide (DMS), methane thiol (CH_3SH), and so on] in wines and beers (Miracle et al., 2005; Herszage and Ebeler, 2011). With these columns the analytes are separated based on differences in their adsorption properties to the stationary phase; the size and shape of the analytes can also influence the separation since the stationary phase particles are porous.

More common for GC separations are gas–liquid separations (GLC) based on partitioning of the analytes between the gas phase and a liquid stationary phase. The column stationary phase can vary in composition and polarity ranging from highly nonpolar (polydimethylsiloxane) to highly polar (polyethylene glycol) phases with a range of intermediate polarity phases also possible. Analytes interact with the phases based on van der Waals forces, dipole–dipole interactions, and hydrogen bonding interactions. Generally, more polar analytes will be more retained on more polar stationary phases and vice versa. The vast majority of GC analyses for alcoholic beverages will involve GLC separations.

Column selection for any given application will depend on the analytes being measured. For many alcoholic beverages, the volatile components are dominated by alcohols and aldehydes which are often optimally separated using a more polar stationary phase. However, there are many exceptions and for complex mixtures it can be difficult to optimize the separation for all components in the mixture.

In addition to the choice of stationary phase, the analyst also has options when selecting the column length, the column diameter, and the amount of liquid phase. In general, longer columns give better resolution (i.e., separation) of closely eluting analytes, however, analysis times increase proportionally to the column length (i.e., doubling the column length will double the analysis time but will only improve resolution by ~1.4 times). Narrower diameter columns are better able to resolve closely eluting peaks, but high gas mobile phase pressures are needed and the narrow diameters may limit the amount of sample capacity that can be introduced onto the column. Narrow columns usually are coated with thinner films of stationary phase, which can enhance mass transfer between the liquid phase and the gas phase (resulting in enhanced resolving power) but also limits sample capacity. Most commercially available GC columns are made of narrow bore fused silica with a polyimide coating to provide strength and flexibility to the column. Column inner diameters range from ~0.2 to 0.53 mm with lengths of 10–60 m. The liquid phase is coated on the inner surface of the column and is usually chemically bonded to the surface. Liquid phase thicknesses range from 0.1 to 5 μm. Column manufacturers provide useful information on column selection for a variety of applications.

Common universal detectors for GC are flame ionization detector (FID) and MS. With an FID, the analytes burn in a flame as they elute from the column. Analytes that contain reduced carbon (i.e., —CH—) burn in the flame and become ionized to $—CHO^+$ (plus an electron) which is collected at a negatively charged collector; the change in current is converted to a signal that is proportional to the analyte concentration (Klee, 2012). The FID is considered a "universal" detector responding to all organic analytes and is widely used for alcoholic beverage analysis. With the development of relatively cheap table-top mass spectrometers that are easy to use and which give additional information about analyte mass and structure, most analytical laboratories now employ GC-MS analysis for many applications. Selective detectors such as electron capture detectors (ECD), nitrogen-phosphorous detectors (NPD), and sulfur specific detectors [e.g., flame photometric detectors (FPD), sulfur chemiluminescence detectors (SCD)] respond specifically to halogen-, nitrogen- or phosphorous-, and sulfur-containing analytes, respectively. Manufacturers provide detailed information about use and applications of these detectors.

Multidimensional GC separations, including "heart-cutting" where a peak or section of the chromatogram is isolated from the analytical column and transferred to a different column (and sometimes to a second GC instrument), provide separation of coeluting peaks. Heart-cutting has been especially used for analysis of chiral compounds where the first column separates the analytes based on sorption and/or partitioning characteristics and the second column is a chiral column that will separate the enantiomers. Luan et al. (2004, 2006) provide an excellent example of multidimensional heart-cutting for analysis of chiral terpenes in grapes/wines. With comprehensive GC, or GC × GC, all peaks in the chromatogram are transferred to a second column of different polarity or chemistry. Using GC × GC many coeluting peaks can be separated that are not separated on a single column. GC × GC has been used for analysis of trace levels of 3-isobutyl-2-methoxy pyrazine in grapes where an unknown peak interfered with quantitation of the pyrazine (Ryona et al., 2009; Schmarr et al., 2010), to analyze over 350 components in Australian Cabernet Sauvignon wines (Robinson et al., 2011b), and to measure volatiles in vodkas (Wisniewska et al., 2016).

6.2.2 Liquid chromatography

Liquid chromatographic separations were demonstrated in seminal experiments by Tswett (1906a,b). These experiments showed that plant pigments could be separated in a chromatographic column and the separations were based on the chemical interactions of the pigments with calcium carbonate stationary phase, not due to a simple size (molecular weight) filtration of the molecules. Since that time, developments in instrumentation and columns have provided for increasing number of applications for many types of food and beverage analyses. When performed in columns under pressure, high performance liquid chromatography (HPLC), can separate hundreds of compounds simultaneously and can be used for a range of nonvolatile, polar, and nonpolar analytes with molecular weights up to several thousand daltons (e.g., proteins, polysaccharides, tannins). HPLC is widely used for analysis of compounds in alcoholic beverages that influence taste (sugars, organic acids, phenolics), color (phenolics and anthocyanins, carotenoids), and mouthfeel (phenolics and tannins) (Conte et al., 2011; Nollet and Toldrá, 2013).

HPLC stationary phases separate analytes based on adsorption, partitioning, ion exchange, and size exclusion mechanisms. The adsorption and partitioning mechanisms are similar to those described for GC, but the stationary phase chemistries are somewhat different with the most common phases being either alumina, silica gel, or silica particles chemically reacted with a long chain hydrocarbon (e.g., an octadecyl-, C18, group). Ion exchange separations use a resin that is positively or negatively charged at the surface; charged analytes bind to the surface and are retained on the column. The bound analytes are released by changing the pH or ionic strength of the mobile phase. Either cations or anions can be analyzed depending on the ion exchange resin used (Fig. 6.2).

Cation exchange

$$RzX^-H^+ + C^+ \longrightarrow RzX^-C^+ + H^+$$

$X^- = SO_3^-$ (sulfonic acid); COO^- (carboxylic acid)

Anion exchange

$$RzY^+OH^- + A^- \longrightarrow RzY^+A^- + OH^-$$

$Y^+ = -N(R_3)^+$ (quaternary amine); $-NH_3^+$ (amine)

Figure 6.2
Ion exchange reactions for separation of charged analytes
C^+ refers to positively charged cation analytes; A^- refers to negatively charged anion analytes; Rz refers to the ion exchange resin.

Size exclusion (or gel permeation) separations use silica or polymeric stationary phases containing uniform pores. Analytes travel through the column based on their molecular size and shape—small molecules enter in to the stationary phase pores and so travel a greater distance in the column and have longer elution times. Larger molecules flow between the stationary phase particles and elute more quickly.

While HPLC separations may occur at slightly elevated temperatures (usually <80°C), changing column temperature is not the main way to elute analytes from the stationary phase as is done with GC. With HPLC, the mobile phase is a liquid and by changing the composition and flow rate of the mobile phase, analytes that are retained on the column can be eluted. For so-called "normal phase" separations using a polar stationary phase (e.g., silica, alumina, or amino columns), the mobile phase starts out being highly nonpolar (e.g., hexane) and by increasing the mobile phase polarity over time (e.g., to isopropanol, methanol, or water) polar analytes are separated and eluted from the column. With so-called "reverse phase" separations using nonpolar stationary phases (e.g., C18), the starting mobile phase usually contains high percentages of water and/or an aqueous buffer. By changing to increasingly nonpolar mobile phases such as methanol or acetonitrile, more nonpolar analytes elute from the column and are separated. Care must always be taken to ensure that mobile phases are miscible, otherwise pumps used to control the mobile phase flow rate can freeze (e.g., changing mobile phases from hexane to methanol in a normal phase separation would require mixing first with a solvent of intermediate polarity such as acetone that is soluble in both hexane and methanol).

Column length, column diameter, and particle diameter can also influence separations. Typical column lengths range from 50 to 300 mm with column inner diameters of 4.6 mm and particle sizes from 3 to 5 μm. As with GC, longer and narrower columns lead to better separations. Small particle sizes also provide the best resolution. However, particularly with decreasing column diameter and particle size, increased mobile phase pressures are needed to ensure that analytes and mobile phase will flow through the column. In the past 10–15 years, ultrahigh performance liquid chromatography (UHPLC) instrumentation has been developed specifically to handle the very high pressures (up to 1000 bar) and narrow peaks obtained when using short (~300 mm), narrow (~2.1 mm inner diameter) columns with particle diameters of <2 μm. These UHPLC columns and instruments usually result in very fast analysis times (<10 min) with excellent separation power.

One of the most common HPLC/UHPLC detectors is a UV-visible or photodiode array (PDA) detector. The PDA detectors measure the absorption of energy in the UV-visible range as analytes elute from the column. All wavelengths are measured simultaneously using a series of photodiodes and a full UV-visible spectrum of the analyte can be obtained which aids in compound identification. The detector is nondestructive, but not all analytes of interest absorb in the UV/visible region (Section 5.10) and as a result, either a different type of detector must be used or precolumn or postcolumn derivatization of the analytes may be used to form new compounds with UV-visible chromophores. Solvents which strongly absorb in the UV-visible region must be avoided when using UV-visible detectors. Refractive index detectors (Section 5.2) can be used to monitor changes in refractive index of the eluant as analytes elute from the column. Refractive index detectors are commonly used for analysis of sugars in alcoholic beverages (Peris-Tortajada, 2000). As with GC, MS detectors are becoming more commonly used with HPLC/UHPLC and will be discussed in Section 6.3.

Multidimensional separations with HPLC have been used for separations of proteins, peptides, and synthetic polymers (Carr et al., 2012; Wu et al., 2012; Uliyanchenko et al., 2012; Schoenmakers and Aarnoutse, 2014). Multidimensional HPLC separations have not been widely used for alcoholic beverages, however, recently, Willemse et al. (2015) used comprehensive two-dimensional UHPLC to separate anthocyanins and large molecular weight pigments in red wines.

6.3 Mass spectrometry

MS is becoming the detector of choice for a variety of chromatographic separations. A mass spectrometer separates charged analytes based on their mass to charge ratio (m/z). Analytes are ionized as they elute from the chromatographic column. When combined with GC, two ionization techniques are common, chemical ionization (CI) and electron ionization (EI); electron ionization sources predominate. With EI, ionization occurs in a source that is at a reduced pressure and the predominant molecules in the source chamber are the analyte molecules. In the source, analytes are bombarded with high energy electrons (70 eV), loose an electron, become positively charged, and are accelerated to the mass analyzer. The analyte molecules also will fragment into smaller components based on the molecular structure of the initial molecule. The m/z of the molecular ion and fragment ions can be used to identify the original molecular structure.

When combined with HPLC, typical ionization mechanisms are electrospray ionization (ESI) and atmospheric pressure chemical ionization (APCI). With ESI, the HPLC eluent is sprayed through a capillary and across a high potential difference resulting in charged droplets. The spray is heated resulting in evaporation of the solvent in the eluent. Subsequently, as the droplet size decreases, the droplets explode via a Coulombic explosion and single- or multiple-charged analyte ions are formed. There is usually little fragmentation of the analytes, however, the analytes may take on multiple charges depending on their structure and as a result large macromolecules can be analyzed. The analytes may become protonated or deprotonated, $[M + H]^+$ or $[M - H]^-$ respectively, depending on the analyte and the ionizing voltage. With APCI, the HPLC eluent is heated to a high temperature forming aerosols. The analyte molecules in the aerosol are ionized in an electron cloud formed from a corona discharge electrode yielding either protonated or deprotonated molecules. As with ESI, the molecules may be multiple charged, depending on their structure, and minimal fragmentation occurs. Because of the higher source temperature with APCI, compounds that are thermally labile are better analyzed with an ESI source. APCI is also not well suited for analytes that are nonvolatile. The ESI source is better suited for more polar analytes, compared to APCI which can be used for both polar and nonpolar analytes.

Once the analyte is ionized (and fragmented, depending on the source energy and the molecular structure) the positively charged ions are accelerated into the detector where the ions are separated and analyzed based on their m/z. Historically, mass analysis occurred using a magnetic sector instrument. More commonly now, the mass detector will be a transmission quadrupole, a quadrupole ion trap, or a time-of-flight detector. In these instruments, the ions travel through the detector based on their mass to charge ratio and arrive at the detector based on their trajectory through an oscillating electric field or their velocity through a flight tube. It is beyond the scope of this chapter to go into detail on the various types of MS detectors and the reader is referred to recent reviews for more information (Griffiths, 2008; Di Stefano et al., 2012; Himmelsbach, 2012; Ebeler and Mitchell, 2016).

Over the past 10–15 years, additional configurations of hyphenated chromatographic techniques have become widely available commercially. These techniques use multiple types of MS separation to enhance sensitivity, reduce background interference, and improve mass resolution. These hyphenated techniques provide opportunities for highly selective and sensitive targeted analyses where the identity of the analytes is known ahead of time, as well as nontargeted analyses which provide opportunities to profile a large number of compounds in the sample even if their identity is not known (Dunn et al., 2013). When used with high resolution instruments and by matching spectral information to library databases, the unknowns can be identified if desired, but need not be in every case. In these cases, the peaks may be referred to as mass features or unknowns, but statistical data processing and comparisons among samples can still occur. In the following sections we briefly discuss these targeted and nontargeted approaches as applied to alcoholic beverages.

6.3.1 Targeted analysis with selected ion monitoring and tandem mass spectrometry

In the one of the simplest types of targeted MS analysis, a GC-MS or HPLC-MS instrument is used in the "selected ion monitoring" (SIM) mode where the MS is set to only scan for selected ions associated with the analyte/s of interest (instead of scanning across a wide m/z range throughout the entire chromatogram as is typically done). The overall sensitivity for a given analyte is increased because the

MS can collect more ions of the targeted masses within a given time window. SIM was used by Allen in early GC-MS analysis of methoxypyrazines in grapes and wines and allowed for detection limits of 0.2 ng/L, a concentration below the perceived sensory threshold (Harris et al., 1987; Allen and Lacey, 1993; Allen et al., 1994). The ions selected should be unique to the analyte of interest at a given retention time window and the ion intensity should be large enough to give good signal response and sensitivity in the concentration range expected in the samples. Typically, two to three ions will be used for each analyte of interest, with one ion being used for quantitative purposes and the other two ions used as so-called qualifier ions to ensure that the ion ratios in the peak match those of the actual compound. Many new GC instruments allow simultaneous SIM/scan monitoring; here the instrument automatically switches between SIM and scan mode allowing for enhanced sensitivity of trace analytes while obtaining full mass spectral interpretation for peak identification. Hjelmeland et al. (2013) and Hendrickson et al. (2016) describe application of a synchronous SIM/scan method to profile volatiles in Cabernet Sauvignon and Pinot noir wines, in both cases focusing on volatiles associated with wine aroma.

More recently, tandem MS (MS/MS or MS^n) configurations allow for two (or more) MS analyses to occur after the analytes elute from the column. There are two types of MS/MS instruments. In the first configuration, the MS analyzers are placed sequentially in the instrument and analyte ions enter the analyzers one after the other. In the other configuration, the MS analyzer is an "ion trap" that holds all the ions together in the analyzer and manipulates them in time so that only ions of a given mass are released and detected at a specific time. In both MS configurations, a reaction or collision cell is used to induce fragmentation of the analytes. The reaction cell is placed in between two MS analyzers in the case of instruments with sequential mass analyzers. For ion traps, the trap also serves as the collision cell and the collision gas is introduced into the trap. Tandem MS detection can be applied in several ways (1) all analyte ions (precursors) enter the first mass analyzer, MS1, are separated and fragmented in the collision cell and only selected product ions that are formed in the collision cell are monitored and detected in the second mass analyzer, MS2; (2) MS1 allows only selected analyte ions (precursors) to react in the collision cell and all subsequent product ions are separated and detected in MS2; (3) MS1 allows all analyte ions (precursors) to react in the collision cell while MS2 scans only for masses that are associated with a constant neutral loss from the precursor ions; and (4) both MS1 and MS2 select only for specific preidentified masses. In all cases, the use of the two (or more) mass analyzers allows for removal of analytes and ions that are not of interest and so the background baseline levels are very low, resulting in enhanced sensitivity for the analytes of interest. By careful selection of MS1 and MS2, some structural information can be obtained (e.g., identify the product ions detected in MS2 that are formed from specific precursor ions selected in MS1). For more detailed discussions of the theory and applications of tandem MS the reader is referred to several published reviews (Flamini, 2003; Flamini and Panighel, 2006; de Hoffmann and Stroobant, 2007; Herrero et al., 2012; Himmelsbach, 2012).

In grapes and wines, tandem MS has been used for targeted analysis of smoke taint glycosides (Hayasaka et al., 2010a,b; Dungey et al., 2011), to quantify trace levels of compounds that impact wine aroma, such as haloanisoles (Hjelmeland et al., 2012) and methoxypyrazines (Hjelmeland et al., 2016), and to rapidly profile over 150 polyphenols in rosé wines (Lambert et al., 2015). Applications to beer include quantification of over 26 hop-derived bitter compounds (Intelmann et al., 2009), analysis of proanthocyanidin monomers, dimers, and trimers in lager beers (Callemien and Collin, 2008), and trace level analysis of aroma active hop-derived polyfunctional thiols in beer, hop water extract, and wort (Ochiai et al., 2015).

6.3.2 Nontargeted analysis and high resolution MS profiling

In nontargeted analyses, the analytes of interest are not defined ahead of time and all the components in the sample are monitored, with or without identification (Dunn et al., 2013). Quantitation is usually semiquantitative with responses reported relative to an internal standard (see also Section 6.4). For example, Robinson et al. (2011b) used a nontargeted approach to analyze volatiles in Australian wines and to compare volatile composition of wines from different growing regions (Robinson et al., 2011a). Over 350 components were monitored and Retention Indices and mass spectra recorded for tentative identification; several compounds could not be identified, but by reporting the Retention Index (RI) and MS information, consistent with the Metabolomics Standards Initiative (MSI) Chemical Analysis Working Group (CAWG) (Sumner et al., 2007) future studies may be able to identify these components. See also Section 6.4 for further information on compound identification.

High resolution mass spectrometric instruments, such as time-of-flight MS (TOF MS) and ion cyclotron resonance MS (ICR MS) have made nontargeted profiling increasingly common for analysis of alcoholic beverages. The instruments can be hyphenated with GC, HPLC, or used with direct sample introduction and no chromatographic preseparations. These instruments give accurate mass information with mass accuracies of ~10 ppm or better allowing for determination of potential molecular formulas of unknown components in the sample; for a given molecular weight a limited number of compounds will have the same exact mass. Mass accuracy (or mass error) is the difference between the accurate mass and the measured mass. For example, if the monoisotopic calculated exact mass of a molecule is 100.000 Da and the measured accurate mass is 100.001 Da, the difference is 0.001 millimass units (mmu). The error is calculated as [difference/exact mass $\times 10^6$]. For this example, the error is $[(0.001/100) \times 10^6) = 10$ ppm (Webb et al. 2004). In addition, high resolution instruments frequently have the capability of performing tandem MS experiments so that further structural information can be obtained (e.g., combining a quadrupole mass filter with a time-of-flight mass analyzer, qTOF MS).

Nontargeted profiling approaches have been used for analysis of a variety of alcoholic beverages. For example, Gougeon and coworkers used ICR MS to characterize the chemical diversity of wines from different growing regions and wines stored in oak obtained from different forests (Gougeon et al., 2009; Roullier-Gall et al., 2014a,b). Using ICR MS, Liger-Belair et al. (2009) characterized the chemical components in Champagne. They observed that compared to the bulk Champagne, many volatile aroma compounds partitioned preferentially into the aerosol that is created above the sample by the bursting CO_2 bubbles. Collins et al. (2014) used UHPLC qTOF MS to characterize different American whiskey styles (Bourbon, Tennessee, and Rye) and to differentiate whiskeys from different producers and of different ages. Hughey et al. (2016) used UHPLC qTOF MS to characterize components in India Pale Ale beers produced over two production years.

In selected recent applications, information from nontargeted metabolite profiling of grapes and wines has been combined with other –omics analyses. For example, UHPLC qTOF MS profiles were related to transcriptomic profiles to study changes in grape metabolism as a function of *Botrytis cinerea* infection (Noble rot) (Blanco-Ulate et al., 2015) and to differences in the microbiome of Chardonnay and Cabernet Sauvignon grapes and wines from different growing regions in Napa and Sonoma Counties in California (Bokulich et al., 2016). Dal Santo et al. (2013) compared environmental and agronomic impacts on grape transcriptomic and metabolite profiles and showed that many grape genes exhibit phenotypic plasticity and as a result, environment and genotype interactions can have a significant effect on the grape metabolite profiles. In these studies, compounds that have been associated with grape and wine flavor were found in the grapes and corresponding wines (if analyzed). Although no sensory analyses were performed in these studies they do demonstrate the potential power of multiplatform systems biology approaches for understanding changes in chemical and sensory properties as a function of the raw materials of alcoholic beverages (e.g., grapes, hops, and so on), the environmental and ecological variables such as the microbiome, and the processing conditions.

6.4 Compound identification and quantification by GC and HPLC

Compound identification in chromatographic methods is based on comparing retention time and spectral properties with information in databases [e.g., NIST (http://chemdata.nist.gov/mass-spc/ms-search/; NIST/NIH/EPA Mass Spectral Library, 2011), Metlin (https://metlin.scripps.edu/index.php)] or by comparing to retention times and spectra of authentic standards. As retention times vary with the instrument, column conditions, and temperature, a system of identifying relative retention times has been developed for GC. This system, referred to as Kovat's Index (KI) or RI, calculates the retention of an analyte relative to that of a series of alkane hydrocarbons (or alkyl ethyl esters). By convention, hexane with six carbons has a designated KI of 600 while heptane has a KI of 700. A compound that elutes exactly in the middle of these two hydrocarbons would have a KI of 650. Formulas for calculating KI are found in general analytical chemistry textbooks (Harris, 2003; Skoog et al., 2007). Lists of KI values are available online (Acree and Arn, 2004; El-Sayed, 2016; NIST, 2016) and in Jennings and Shibamoto (1980). By matching the KI of an unknown with that in the database, tentative identifications can be obtained. Matching additional spectral information, such as the UV-visible spectrum or mass spectrum, can provide further confirmation of the compound identity. However, until KI and spectral data are matched with authentic standards, ideally on two columns of different polarity, a compound identification is considered tentative (Molyneux and Schieberle, 2007). Large, normalized databases for HPLC have not been published, however, many databases provide retention time information using defined conditions so that retention times of unknowns can be compared.

Quantitation in GC and HPLC is based on peak area or peak height of integrated peaks in the chromatogram. Standard integrators will determine chromatographic baselines based on user input, however, the user is cautioned to carefully monitor the integrations and ensure consistency through a series of sample runs. Deconvolution software can also be used to distinguish overlapping peaks based on similarities and differences in the mass spectrum at different points along a chromatographic peak (Du and Zeisel, 2013).

If standards are available, an external calibration curve can be created by plotting the integrated peak area or peak height of the standards (*y*-axis) against the concentration of the standards (*x*-axis). The regression line from the calibration curve is used to determine the concentration of an unknown based on its integrated peak area or peak height from chromatogram.

Use of internal standards can also aid quantitation. In this case, one or more internal standards (IS) are added to each sample prior to sample preparation and/or prior to injection on the GC. If the IS is added prior to sample preparation it can be used to correct for sample losses during the preparation steps. If the IS is added prior to sample injection, it will account for variability in injection volume and instrument response. The IS should be selected to match as closely as possible the chemical and physical properties of the analytes of interest, but not be present in the samples. Ideal internal standards are stable isotope matched internal standards that are identical to the analytes but have stable isotopes such as deuterium (^{2}D), ^{13}C, or ^{15}N, replacing one or more H, C, or N atoms in the molecule. A calibration curve for each analyte can be created using the IS where the concentration ratio of the standard to the IS is plotted against the peak area (or peak height) ratio of the standard to the IS; the regression line of the calibration curve is used to determine the concentration of the analytes in unknown sample. When a stable isotope IS is used, a MS detector is required to distinguish the peaks of the analyte and the standards based on their differences in mass. Use of stable isotope dilution analysis (SIDA) for quantitation provides for the most accurate quantitation of analytes (Schieberle and Molyneux, 2012) and is widely used for alcoholic beverage analysis. Siebert et al. (2005) show an excellent application where stable isotopes for 29 different analytes are used to quantify 31 different compounds in wine samples. Hjelmeland et al. (2012) describe a method where stable isotopes were available for only selected haloanisole analytes in wine samples, and so the IS most closely matching the structure of the analyte was used for each haloanisole species quantified. Finally, Canuti et al. (2009) describe a method for analysis of grape volatiles using two different ISs. One was added at the beginning of sample preparation and one prior to sample injection on the GC. Neither IS was a stable isotope matched compound, but both were alcohols selected to provide functional properties similar to compounds in the sample. In cases where authentic standards were not available and calibration curves could not be created, the final concentration of the measured analytes was reported relative to the IS concentration.

6.5 Sample preparation for GC and HPLC

Careful sampling and sample preparation are critical for accurate and precise measurements for any analytical method. Greenfield and Southgate (2003) provide detailed information on obtaining representative food and beverage samples for analytical testing. The analyst should also consider how the samples are handled and stored between sample collection and analysis as well as appropriate record keeping protocols.

In most cases, the sample will require further preparation after sample collection and prior to analysis. This may include steps like grinding/homogenization to break up solid samples such as grapes or hops, filtration of liquids to remove particulates, dilution to decrease analyte concentrations to within the range of instrumental detection, and/or drying to remove moisture if analytes are to be compared on a dry-weight basis. Even relatively simple steps like these however, can cause analyte losses (e.g., oxidation from exposure to light or air during homogenization or grinding, sorption or binding to filter materials, and volatilization of compounds) so the analyst should always be aware of the impact of any sample preparation steps on analyte losses and/or chemical changes. Analyte recoveries can be determined and internal standards can also be added to correct for these losses (see also Section 6.4).

Other sample preparation steps may include distillation to remove nonvolatiles for GC analysis and/or to concentrate the analytes of interest. The distillation may be combined with solvent extraction and a number of devices have been developed for simultaneous distillation/extraction particularly for GC sample preparation. Jennings and Rapp (1983) and Majors (2013) provide reviews of these methods. These distillation/extraction methods are often time consuming, require large amounts of solvents, and depending on the actual solvents and temperatures used for distillation and extraction, the high temperatures can result in artifactual changes to the sample.

In recent years, sorptive extraction techniques, such as solid phase extraction (SPE) have been widely used to remove matrix components and concentrate the sample. SPE is a low resolution liquid chromatography separation with a stationary phase (e.g., C18) packed into a small cartridge (e.g., 3–6 mL cartridge volumes are common with 0.5–1.0 g of packing material) or in a 96 well plate. The sample is loaded onto the cartridge, analytes bind to the stationary phase, and are then eluted with a small amount of a solvent. A range of stationary phases are available and manufacturers provide detailed information on procedures and applications for SPE.

Solid phase microextraction (SPME), stir bar sorptive extraction (SBSE), and headspace sorptive extractions (HSSE) all use a polymeric phase that is coated onto the surface of either a 1–2 cm long fused silica capillary in the case of SPME or a magnetic stir bar in the case of SBSE and HSSE. A range of nonpolar, midpolarity, polar, and mixed mode phases are available for SPME applications. For SBSE and HSSE, the most common phase is polydimethylsiloxane although polar polyethylene glycol phases are also now commercially available. SPME sampling can occur in either the headspace or the liquid phase, SBSE is used for extraction of analytes from the liquid phase, and HSSE is used for extraction of analytes from the headspace above a sample. In all cases, analytes adsorb to or partition into the polymeric phase from the liquid or vapor phase of the sample and the amount of sorption will depend on the analyte, the chemistry of the polymeric phase, the temperature, the sampling time, and the partition coefficients for the analyte between the liquid, vapor, and polymer phases. Compared to SPME, SBSE, and HSSE have more sorptive phase and so in general have greater sample capacity, however, longer extraction times may be required with SBSE and HSSE for accurate and reproducible extractions. SPME is relatively rapid and most SPME analyses use extraction times of less than 60 min. SPME in particular has become widely used for analysis of volatiles in alcoholic beverages due to its simplicity and ability to analyze trace compounds. Detailed discussions of these techniques are beyond the scope of this chapter and the analyst is referred to reviews by Risticevic et al. (2010a,b), Ebeler (2012), Majors (2013), and Robinson et al. (2014).

Derivatization reactions are also used in sample preparation as noted in Section 6.2.1 of this chapter. The derivatization can improve chromatographic behavior as noted for fatty acids or it can add a functional group or chromophore that will enhance detectability or increase detection selectivity (e.g., add a nitrogen or sulfur group for selective detection by a nitrogen or sulfur GC specific detector). Knapp (1979) and Blau and King (1978) provide detailed information on derivatization reactions for chromatography and Busch (2010) provides a more recent review on applications for hyphenated MS techniques. Careful consideration of derivatization efficiency and analyte recoveries/losses should be considered when performing derivatization reactions.

6.6 Elemental analysis and ICP-MS

Inorganic elements are important plant and yeast nutrients, they can influence chemical reactions and stability of alcoholic beverages, and they can impact human health if present at high levels. Atomic absorption and atomic emission spectroscopy have been widely used to measure elements but they can usually be used for only one or a small number of elements at a time (Harris, 2003; Skoog et al., 2007). Recently inductively coupled plasma (ICP) MS has been used for elemental analysis of alcoholic beverages. The high temperature argon plasma efficiently ionizes most elements (Fig. 6.3) and the MS can then detect them based on their masses. As a result, ICP MS is ideal for simultaneously profiling numerous trace elements. Elements present in high concentrations can be analyzed by microwave plasma atomic emission detection (MP AES). Elements are excited in the microwave plasma and as the excited element drops to a lower energy state the emission of light is monitored at wavelengths characteristic of the elements measured, similar to light emission with atomic emission spectroscopy. However, compared to flame atomic absorption and atomic emission spectroscopy, the microwave plasma temperature is higher and so excitation of individual elements is improved resulting in lower detection limits. MP AES instruments are overall less expensive than ICP MS instruments and the nitrogen plasma is cheaper and easier to use than an argon plasma used for ICP-MS. Georgiou and Danezis (2015) provide a recent review of elemental analysis of foods and beverages.

Example applications of elemental analysis for alcoholic beverages include distinguishing tequila and mezcal samples (Ceballos-Magaña et al., 2009); discrimination of beer styles from different manufacturers (Mahmood et al., 2012); evaluation of the effects of packaging and storage temperature on metal content in wines (Hopfer et al., 2013); monitoring elemental composition in grapes and wines as a function of growing region and winemaking conditions (Hopfer et al., 2015; Nelson et al., 2015); and monitoring water quality for beer making (Bamforth, 2000; Eumann and Schaeberle, 2016).

In a final application of elemental analysis, we mention isotope ratio mass spectrometry (IRMS). Here, samples are combusted and the ratios of stable isotopes of one or several elements are measured (e.g., $^{2}H/^{1}H$; $^{18}O/^{16}O$; $^{13}C/^{12}C$; $^{15}N/^{14}N$). The most common applications are for determination of adulteration.

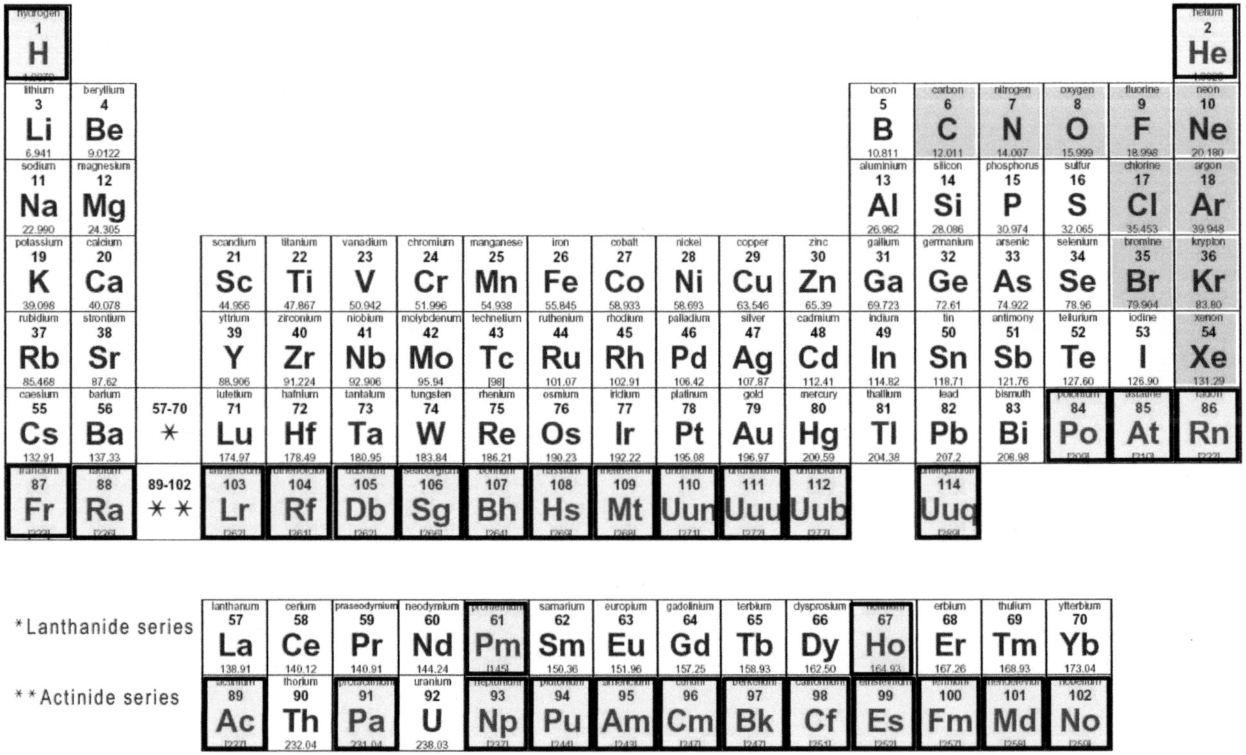

hydrogen 1 H																		helium 2 He
lithium 3 Li 6.941	beryllium 4 Be 9.0122												boron 5 B 10.811	carbon 6 C 12.011	nitrogen 7 N 14.007	oxygen 8 O 15.999	fluorine 9 F 18.998	neon 10 Ne 20.180
sodium 11 Na 22.990	magnesium 12 Mg 24.305												aluminium 13 Al 26.982	silicon 14 Si 28.086	phosphorus 15 P 30.974	sulfur 16 S 32.065	chlorine 17 Cl 35.453	argon 18 Ar 39.948
potassium 19 K 39.098	calcium 20 Ca 40.078		scandium 21 Sc 44.956	titanium 22 Ti 47.867	vanadium 23 V 50.942	chromium 24 Cr 51.996	manganese 25 Mn 54.938	iron 26 Fe 55.845	cobalt 27 Co 58.933	nickel 28 Ni 58.693	copper 29 Cu 63.546	zinc 30 Zn 65.39	gallium 31 Ga 69.723	germanium 32 Ge 72.61	arsenic 33 As 74.922	selenium 34 Se 78.96	bromine 35 Br 79.904	krypton 36 Kr 83.80
rubidium 37 Rb 85.468	strontium 38 Sr 87.62		yttrium 39 Y 88.906	zirconium 40 Zr 91.224	niobium 41 Nb 92.906	molybdenum 42 Mo 95.94	technetium 43 Tc [98]	ruthenium 44 Ru 101.07	rhodium 45 Rh 102.91	palladium 46 Pd 106.42	silver 47 Ag 107.87	cadmium 48 Cd 112.41	indium 49 In 114.82	tin 50 Sn 118.71	antimony 51 Sb 121.76	tellurium 52 Te 127.60	iodine 53 I 126.90	xenon 54 Xe 131.29
caesium 55 Cs 132.91	barium 56 Ba 137.33	57-70 *	lutetium 71 Lu 174.97	hafnium 72 Hf 178.49	tantalum 73 Ta 180.95	tungsten 74 W 183.84	rhenium 75 Re 186.21	osmium 76 Os 190.23	iridium 77 Ir 192.22	platinum 78 Pt 195.08	gold 79 Au 196.97	mercury 80 Hg 200.59	thallium 81 Tl 204.38	lead 82 Pb 207.2	bismuth 83 Bi 208.98	84 Po [209]	85 At [210]	86 Rn [222]
87 Fr [223]	88 Ra [226]	89-102 **	103 Lr [262]	104 Rf [261]	105 Db [262]	106 Sg [266]	107 Bh [264]	108 Hs [269]	109 Mt [268]	110 Uun [271]	111 Uuu [272]	112 Uub [277]		114 Uuq [289]				

*Lanthanide series	lanthanum 57 La 138.91	cerium 58 Ce 140.12	praseodymium 59 Pr 140.91	neodymium 60 Nd 144.24	61 Pm [145]	samarium 62 Sm 150.36	europium 63 Eu 151.96	gadolinium 64 Gd 157.25	terbium 65 Tb 158.93	dysprosium 66 Dy 162.50	67 Ho 164.93	erbium 68 Er 167.26	thulium 69 Tm 168.93	ytterbium 70 Yb 173.04
**Actinide series	89 Ac [227]	thorium 90 Th 232.04	91 Pa 231.04	uranium 92 U 238.03	93 Np [237]	94 Pu [244]	95 Am [243]	96 Cm [247]	97 Bk [247]	98 Cf [251]	99 Es [252]	100 Fm [257]	101 Md [258]	102 No [259]

Figure 6.3
Periodic table of elements
Elements in gray have relatively high first ionization efficiencies and are not readily ionized in an argon plasma used for ICP-MS (5000K, ionization potential greater than 11 eV; Houk 1986). Elements in light gray with a heavy outline are not usually measured by ICP-MS.

For example, measurement of $^{13}C/^{12}C$ and $^{18}O/^{16}O$ ratios has been used to determine if tequila has been fermented with sugars from the agave plant or from other sugars based on differences in carbon fixation of the plant sources (Aguilar-Cisneros et al., 2002). Illegal watering of wines can be detected by measuring ratios of $^{2}H/^{1}H$ and $^{18}O/^{16}O$ (Calderone and Guillou, 2008). $^{13}C/^{12}C$ ratios of CO_2 have been proposed as a way to determine if carbon dioxide in sparkling wines and beer is from natural fermentation processes or from addition of CO_2 gas of fossil origin, however more work is needed to fully explore the ratios obtained from a variety of products, producers, and production methods (Calderone et al., 2007).

6.7 Capillary electrophoresis

Electrophoresis is the differential movement of charged particles in an electric field. In CE, the separation occurs in a narrow-bore fused silica capillary (25–75 μm inner diameter). The capillary is filled with a buffer and the sample, the ends of the capillary are placed in reservoirs containing the same, or a similar buffer, and an electric field is applied between the two reservoirs to attract the charged molecules toward the detector. UV-visible detectors are commonly used, but electrochemical, refractive index, and MS detectors are also possible. A more detailed discussion of theory and application of capillary electrophoresis is found in Frazier et al. (2000) and Harstad et al. (2016).

A variety of alcoholic beverage components can be analyzed by capillary electrophoresis including organic acids, phenolics, biogenic amines, ethanol, sugars, amino acids, isoalpha acids, aromatic aldehydes, vitamins and inorganic ions (Haumann et al., 2000; Panossian et al., 2001; Cortacero-Ramírez et al., 2003, 2007; Wang et al., 2003; Peng et al., 2004; Stremel Azevedo et al., 2014; Oliver et al., 2014; Daniel et al., 2015).

An advantage of CE compared to chromatographic methods is that injection volumes are small (a few nanoliters) which may be important if there is limited sample volume. In addition, CE separations are often very fast (a few seconds to a few minutes), and a variety of analytes (polar, nonpolar, volatile, nonvolatile) can be analyzed with the same basic instrument.

6.8 NMR

NMR spectroscopy is used to monitor chemical and physical properties of atoms in molecules. In most cases, proton (^{1}H), carbon (^{13}C), or oxygen (^{17}O) atoms are monitored and the information can be used to profile the composition of complex mixtures, and to quantify targeted analytes in a sample. Overviews of NMR principles and instrumentation are available in analytical chemistry textbooks (Harris, 2003; Skoog et al., 2007).

NMR has been used for isotopic ratio analysis of alcoholic beverages; as with isotopic ratio mass spectrometry (IRMS), ratios of $^{2}H/^{1}H$ and $^{13}C/^{12}C$ can be used to determine the sugar sources for ethanol fermentation. The $^{2}H/^{1}H$ and $^{18}O/^{16}O$ ratio can also vary depending on the growing climate of the grapes, therefore, the $^{2}H/^{1}H$ ratio determined by NMR combined with $^{18}O/^{16}O$ ratios determined by IRMS has been used for authentication of the geographic origin of grapes and corresponding wines. Christoph et al. (2007) provide an excellent overview of these applications.

NMR metabolite profiling has also been used for authentication and to relate composition and sensory properties. Fotakis et al. (2013a,b) demonstrate the application of NMR profiling to discriminate wines based on vintage, region of origin, and grape cultivar and to discriminate grape marc distillates from different growing regions in Greece. In an interesting application, Skogerson et al. (2009) showed that ^{1}H NMR profiles were able to predict wine mouthfeel properties; proline and lactate in particular were associated with a more viscous wine mouthfeel however it is unclear if these compounds themselves directly contribute to the mouthfeel perception or whether they are correlative markers only.

NMR profiling of alcoholic beverages usually requires minimal sample preparation, however, the detection limits for individual components are usually higher than for GC-MS methods. The NMR metabolites that are typically observed in alcoholic beverages are amino acids, phenolics/anthocyanins, organic acids, sugars, sugar alcohols, and volatile alcohols and esters (Skogerson et al., 2009; Fotakis et al., 2013a,b). As noted by Skogerson et al. (2009) it can be difficult to assign compound identities to NMR spectra of complex mixtures due to significant peak overlap and to limitations in existing NMR metabolite libraries. Due to their high costs, access to NMR instruments for most alcoholic beverage analysis labs may be limited and NMR analyses as described here are usually performed in research settings or by larger analytical and testing laboratories.

6.9 Integrating chemical and sensory analyses

Since the focus of this book is chemical and sensory analysis of alcoholic beverages, we close this chapter with a brief discussion of analytical and statistical approaches for directly relating sensory and chemical properties of foods and beverages. Using these approaches scientists are gaining increased insight into the compounds that directly and indirectly contribute to sensory properties and an improved understanding of how mixtures of compounds are perceived.

GC-olfactometry uses the human nose as a detector for a GC separation. An individual sniffs the effluent from the end of the column and provides a qualitative description of the aromas as they elute. Quantitative intensity measures may also be given. Alternatively, serial dilutions of the samples are analyzed and those compounds which are still sensed after several dilutions are considered to be most important in contributing to the sensory aroma of the original sample. Different approaches for GC-O have been used in the literature including CHARM analysis (Acree et al., 1984; Acree, 1997), Aroma Extract Dilution Analysis (AEDA; Ullrich and Grosch, 1987), OSME (McDaniel et al., 1990), and nasal impact frequency/surface nasal impact frequency analysis (NIF/SNIF; Pollien et al., 1997, 1999). Chin and Marriott (2015) and Chambers and Koppel (2013) have provided recent reviews of GC-O applications for aroma analysis.

A related method, *taste dilution analysis* (*TDA*), has been used by Hoffmann and coworkers (Frank et al., 2001) to identify compounds which contribute to taste and mouthfeel properties of foods and beverages. For Taste Dilution assays, sample extracts are separated into fractions by HPLC analysis (or other fractionation methods) and each fraction is tasted and compared to a control/water sample. Only those fractions which are sensed after several dilutions are considered to contribute to the overall taste or mouthfeel properties. Each fraction may contain one or a small number of components and additional fractionation may occur to yield individual compounds with taste/mouthfeel activity.

Both GC-O and TDA are separatory methods and do not measure the aroma or taste properties of mixtures of compounds. Further, flavor qualities of individual compounds can change with concentration and threshold measurements do not necessarily predict sensory properties at suprathresold levels. Finally, individuals will have variable sensitivities to individual compounds and so careful training and standardization are needed to ensure accurate and precise evaluations across judges and replications. To fully evaluate the sensory properties of aroma or taste active compounds they must be studied in reconstitution or omission mixtures, where the compounds identified by GC-O or TDA are added in concentrations similar to those of the original sample and to a matrix similar to that of the original sample. The sensory properties of the mixtures are then compared to the original sample, and individual compounds added or removed from the mixture to determine which compounds result in mixtures with aroma or taste properties most similar to those of the original sample. The taste or aroma active compounds need to be identified, quantified and isolated, synthesized, or purchased to use in the reconstitution/omission studies.

Recently, Johnson et al. (2012) described a variation of GC-O where the volatiles are mixed together on-column and the effluent smelled as a mixture. By using a switching valve, individual compounds or groups of compounds can be removed from the mixture and the impacts on the aroma of the resulting mixture evaluated. Using this technique, the compounds need not be identified to evaluate the aroma properties of the mixtures, and the concentrations already resemble those found in the original sample so that further quantitation is not needed.

Odor activity values (*OAV*) provide another approach to identifying those compounds which contribute to aroma of complex mixtures. To determine OAVs, compounds in a GC chromatogram are identified and quantified and then aroma thresholds are determined separately in matrices resembling those of the samples of interest (reviewed by Audouin et al., 2001). The OAV is calculated by dividing the concentration of the compound in the sample by the aroma threshold. Compounds with high OAV are considered to contribute more to the overall aroma of the original sample. Limitations of OAV determinations are similar to those of GC-O measurements and have been discussed by Audouin et al. (2001). Published threshold values also are commonly used when determining OAVs but the matrices for published OAVs are usually different than those of the samples of interest and frequently the published thresholds do not specify the type of threshold determined (aroma/orthonasal, retronasal, detection, recognition, etc.).

Finally, chemical and sensory profiles can be related using *multivariate statistical approaches* (i.e., chemometrics or sensometrics). Example applications are described throughout this book and Case Study 10 provides a worked example showing how chemical data can be used to predict sensory properties of wine. These approaches measure sensory and chemical properties of mixtures and so take into account masking, additive, and synergistic effects on sensory perception (Hopfer et al., 2012). However, the analyst is cautioned that the resulting associations/correlations do not necessarily describe cause and effect relationships in all cases.

6.10 Conclusions

Numerous instrumental methods can be used to provide important compositional information for alcoholic beverages. The specific choice of instrumental method will depend largely on the analytes of interest and whether qualitative or quantitative information (or both) is needed to meet the analytical objectives. Additional considerations, including sample preparation requirements, sample sizes, speed of analysis, and cost of instrumentation and operator training should also be considered by the analyst. Although many of the methods described may not be used in smaller production facilities, many large- to medium-sized facilities, as well as regulatory and research laboratories will employ one or more of these instrumental techniques for compositional analysis of alcoholic beverages. Therefore, understanding the basic principles and applications of these approaches can be critical for the analyst interested in ensuring the production of consistent, stable alcoholic beverages that conform to regulatory requirements and that have the desired sensory properties.

References

Acree, T.E., 1997. GC-olfactometry. GC with a sense of smell. Anal. Chem. 69, 170A–175A.

Acree, T., Arn, H., 2004. Flavornet and human odor space. Available from: http://www.flavornet.org

Acree, T.E., Barnard, J., Cunningham, D.G., 1984. A procedure for the sensory analysis of gas chromatographic effluents. Food Chem. 14, 273–286.

Aguilar-Cisneros, B.O., López, M.G., Richling, E., Heckel, F., Schreier, P., 2002. Tequila authenticity assessment by headspace SPME-HRCG-IRMS analysis of $^{13}C/^{12}C$ and $^{18}O/^{16}O$ ratios of ethanol. J. Agric. Food Chem. 50, 7520–7523.

Allen, M.S., Lacey, M.J., 1993. Methoxypyrazine grape flavour: influence of climate, cultivar and viticulture. Die Wein-Wissenschaft 48, 211–213.

Allen, M.S., Lacey, M.J., Boyd, S., 1994. Determination of methoxypyrazines in red wines by stable isotope dilution gas chromatography-mass spectrometry. J. Agric. Food Chem. 42, 1734–1738.

Association of Official Analytical Chemists (AOAC)., 2012. Official Methods of Analysis of AOAC International, nineteenth ed. AOAC International, Gaithersburg, MD.

Audouin, V., Bonnet, F., Vickers, Z.M., Reineccius, G.A., 2001. Limitations in the use of odor activity values to determine important odorants in food. In: Leland, J., Scheiberle, P., Buettner, A., Acree, T.E. (Eds.), Gas Chromatography-Olfactometry. The State of the Art. ACS Symposium Series No. 782. American Chemical Society, Washington, DC, pp. 156–171.

Bamforth, C.W., 2000. Beer: an ancient yet modern biotechnology. Chem. Educator 5, 102–112.

Bamforth, C.W., 2016. Brewing Materials and Processes: A Practical Approach to Beer Excellence. Academic Press, London.

Blanco-Ulate, B., Amrine, K.C.H., Collins, T.S., Rivero, R.M., Vicente, A.R., Morales-Cruz, A., Doyle, C.L., Ye, Z., Allen, G., Heymann, H., Ebeler, S.E., Cantu, D., 2015. Developmental and metabolic plasticity of white-skinned grape berries in response to *Botrytis cinerea* during noble rot. Plant Physiol. 169 (4), 2422–2443.

Blau, K., King, G.S., 1978. Handbook of Derivatives for Chromatography. Heyden, New York, NY.

Bokulich, N., Collins, T., Masarweh, C., Allen, G., Heymann, H., Ebeler, S.E., Mills, D.A., 2016. Associations among wine grape microbiome, metabolome, and fermentation behavior suggest microbial contribution to regional wine characteristics. mBio 7 (3), e00631-16.

Busch, K.L., 2010. Derivatization in mass spectrometry, Spectroscopy, November, 201. Available from: http://www.spectroscopyonline.com/derivatization-mass-spectrometry?id=&pageID=1&sk=&date

Calderone, G., Guillou, C., 2008. Analysis of isotopic ratios for the detection of illegal watering of beverages. Food Chem. 106, 1399–1405.

Calderone, G., Guillou, C., Reniero, F., Naulet, N., 2007. Helping to authenticate sparkling drinks with $^{13}C/^{12}C$ of CO_2 by gas chromatography-isotope ratio mass spectrometry. Food Res. Int. 40 (3), 324–331.

Callemien, D., Collin, S., 2008. Use of RP-HPLC-ESI(-)-MS/MS to differentiate various proanthocyanidin isomers in lager beer extracts. J. Am. Soc. Brew. Chem. 66, 1–7.

Canuti, V., Conversano, M., Li Calzi, M., Heymann, H., Matthews, M.A., Ebeler, S.E., 2009. Headspace solid-phase microextraction-gas chromatography-mass spectrometry for profiling free volatile compounds in Cabernet Sauvignon grapes and wines. J. Chromatogr. A 1216, 3012–3022.

Carr, P.W., Davis, J.M., Rutan, S.C., Stoll, D.R., 2012. Principles of online comprehensive multidimensional liquid chromatography. Grushka, E., Grinberg, N. (Eds.), Advances in Chromatography, 50, CRC Press, Boca Raton, FL, pp. 139–236.

Ceballos-Magaña, S.G., Jurado, J.M., Martín, M.J., Pablos, F., 2009. Quantitation of twelve metals in tequila and mezcal spirits as authenticity parameters. J. Agric. Food Chem. 57, 1372–1376.

Chambers, E., Koppel, K., 2013. Associations of volatile compounds with sensory aroma and flavor: the complex nature of flavor. Molecules 18, 4887–4905.

Chin, S.-T., Marriott, P.J., 2015. Review of the role and methodology of high resolution approaches in aroma analysis. Anal. Chim. Acta 854, 1–12.

Christoph, N., Rossmann, A., Schlicht, C., Voerkelius, S., 2007. Wine authentication using stable isotope ratio analysis: significance of geographic origin, climate, and viticultural parameters.

In: Ebeler, S.E., Takeoka, G.R., Winterhalter, P. (Eds.), Authentication of Food and Wine. ACS Symposium Series No. 952. American Chemical Society, Washington, DC, pp. 166–179.

Collins, T.S., Zweigenbaum, J., Ebeler, S.E., 2014. Profiling of nonvolatiles in whiskeys using ultra high pressure liquid chromatography quadrupole time-of-flight mass spectrometry (UHPLC-QTOF MS). Food Chem. 163, 186–196.

Conte, L.S., Moret, S., Purcaro, G., 2011. HPLC in food analysis. In: Corradini, D. (Ed.), Handbook of HPLC. CRC Press, Boca Raton, FL.

Cortacero-Ramírez, S., Arráez-Román, D., Segura-Carretero, A., Fernández-Gutiérrez, A., 2007. Determination of biogenic amines in beers and brewing-process samples by capillary electrophoresis coupled to laser-induced fluorescence detection. Food Chem. 199, 383–389.

Cortacero-Ramírez, S., Hernáinz-Bermúdez de Castro, M., Segura-Carretero, A., Cruces-Blanco, C., Fernández-Gutiérrez, A., 2003. Analysis of beer components by capillary electrophoretic methods. Tr. Anal. Chem. 22, 440–455.

Dal Santo, S., Tornielli, G.B., Zenoni, S., Fasoli, M., Farina, L., Anesi, A., Guzzo, F., Delledonne, M., Pezzotti, M., 2013. The plasticity of the grapevine berry transcriptome. Genome Biol. 14, 454.

Daniel, D., Bezerra dos Santos, V., Vidal, D.T.R., do Lago, C.L., 2015. Determination of biogenic amines in beer and wine by capillary electrophoresis-tandem mass spectrometry. J. Chromatogr. A 1416, 121–128.

de Hoffmann, E., Stroobant, V., 2007. Mass Spectrometry: Principles and Applications, third ed. Wiley and Sons, West Sussex, England.

Dettmer-Wilde, K., Engewald, W., 2014. Practical Gas Chromatography: A Comprehensive Reference. Springer-Verlag, Berlin.

Di Stefano, V., Avellone, G., Bongiorno, D., Cunsolo, V., Muccilli, V., Sforza, S., Dossena, A., Drahos, L., Vékey, K., 2012. Applications of liquid chromatography-mass spectrometry for food analysis. J. Chromatogr. A 1259, 74–85.

Du, X., Zeisel, S.H., 2013. Spectral deconvolution for gas chromatography mass spectrometry-based metabolomics: current status and future perspectives. Comput. Struct. Biotechnol. J. 4, e201301013.

Dungey, K.A., Hayasaka, Y., Wilkinson, K.L., 2011. Quantitative analysis of glycoconjugate precursors of guaiacol in smoke-affected grapes using liquid chromatography-tandem mass spectrometry based stable isotope dilution analysis. Food Chem. 126, 801–806.

Dunn, W.B., Erban, A., Weber, R.J.M., Creek, D.J., Brown, M., Breitling, R., Hankemeier, T., Goodacre, R., Neumann, S., Kopka, J., Viant, M.R., 2013. Mass appeal: metabolite identification in mass spectrometry-focused untargeted metabolomics. Metabolomics 9 (Suppl. 1), 44–66.

Ebeler, S.E., 2012. Gas chromatographic analysis of wines: current applications and future trends. In: Poole, C.F. (Ed.), Gas Chromatography. Elsevier, New York, NY, pp. 689–710.

Ebeler, S.E., Mitchell, A.E., 2016. Analytical chemistry of botanical extracts. In: Sivamani, R.K., Jagdeo, J., Elsner, P., Maibach, H.I. (Eds.), Cosmeceuticals and Active Cosmetics. third ed. CRC Press, Boca Raton, FL, pp. 411–428, (Chapter 33).

El-Sayed, A.M., 2016. The pherobase: database of pheromones and semiochemicals. Available from: http://www.pherobase.com

Ettre, L.S., 2008. The beginnings of gas adsorption chromatography 60 years ago. LCGC North Am. 26, 48–51.

Eumann, M., Schaeberle, C., 2016. Water. In: Bamforth, C.W. (Ed.), Brewing Materials and Processes: A Practical Approach to Beer Excellence. Academic Press, London, pp. 97–111.

Flamini, R., 2003. Mass spectrometry in grape and wine chemistry. Part I: polyphenols. Mass Spectrom. Rev. 22 (4), 218–250.

Flamini, R., Panighel, A., 2006. Mass spectrometry in grape and wine chemistry. Part II: the consumer protection. Mass Spectrom. Rev. 25 (5), 741–774.

Fotakis, C., Christodouleas, D., Kokkotou, K., Zervou, M., Zoumpoulakis, P., Moulos, P., Liouni, M., Calokerinos, A., 2013a. NMR metabolite profiling of Greek grape marc spirits. Food Chem. 138, 1837–1846.

Fotakis, C., Kokkotou, K., Zoumpoulakis, P., Zervou, M., 2013b. NMR metabolite fingerprinting in grape derived products: an overview. Food Res. Int. 54 (1), 1184–1194.

Frank, O., Ottinger, H., Hofmann, T., 2001. Characterization of an intense bitter-tasting 1H,4H-quinolizinium-7-olate by application of the taste dilution analysis, a novel bioassay for the screening and identification of taste-active compounds in foods. J. Agric. Food Chem. 49, 231–238.

Frazier, R.A., Ames, J.A., Nursten, H.E., 2000. Capillary Electrophoresis for Food Analysis. Method Development. The Royal Society of Chemistry, Cambridge, UK.

Georgiou, C.A., Danezis, G., 2015. Elemental and isotopic mass spectrometry. Adv. Mass Spectrom. Food Saf. Qual. 68, 131–243.

Gougeon, R.D., Lucio, M., Fommberger, M., Peyron, D., Chassagne, D., Alexandre, H., Feuillat, F., Voilley, A., Cayot, P., Gebefügi, I., Hertkorn, N., Schmitt-Kopplin, P., 2009. The chemodiversity of wines can reveal a metabologeography expression of cooperage oak wood. PNAS 106, 9174–9179.

Greenfield, H., Southgate, D.A.T., 2003. Sampling. Food Composition Data. Food and Agriculture Organizations of the United Nations, Rome, (Chapter 5).

Griffiths, J., 2008. A brief history of mass spectrometry. Anal. Chem. 80, 5678–5683.

Harris, D.C., 2003. Quantitative Chemical Analysis, sixth ed W. H. Freeman, New York.

Harris, R.L.N., Lacey, M.J., Brown, W.V., Allen, M.S., 1987. Determination of 2-methoxy-3-alkylpyrazines in wine by gas chromatography/mass spectrometry. Vitis 26, 201–207.

Harstad, R.K., Johnson, A.C., Weisenberger, M.M., Bowser, M.T., 2016. Capillary electrophoresis. Anal. Chem. 88 (1), 299–319.

Haumann, I., Boden, J., Mainka, A., Jegle, U., 2000. Simultaneous determination of inorganic anions and cations by capillary electrophoresis with indirect UV detection. J. Chromatogr. A 895, 269–277.

Hayasaka, Y., Baldock, G.A., Pardon, K.H., Jeffrey, D.W., Herderich, M.J., 2010a. Investigation into the formation of guaiacol conjugates in berries and leaves of grapevine *Vitis vinifera* L. Cv. Cabernet Sauvignon using stable isotope tracers combined with HPLC–MS and MS/MS analysis. J. Agric. Food Chem. 58, 2076–2081.

Hayasaka, Y., Dungey, K.A., Baldock, G.A., Kennison, K.R., Wilkinson, K.L., 2010b. Identification of a β-D-glucopyranoside precursor to guaiacol in grape juice following grapevine exposure to smoke. Anal. Chim. Acta 660, 143–148.

Hendrickson, D.A., Lerno, L.A., Hjelmeland, A.K., Ebeler, S.E., Heymann, H., Hopfer, H., Block, K.L., Brenneman, C.A., Oberholster, A., 2016. Impact of mechanical harvesting and optical berry sorting on grape and wine composition. Am. J. Enol. Vitic., [Epub ahead of print].

Herrero, M., Simó, C., García-Cañas, V., Ibáñez, E., Cifuentes, A., 2012. Foodomics: MS-based strategies in modern food science and nutrition. Mass Spectrom. Rev. 31 (1), 49–69.

Herszage, J., Ebeler, S.E., 2011. Analysis of volatile organic sulfur compounds in wine using headspace-SPME-GC with sulfur chemiluminescence detection. Am. J. Enol. Vitic. 62, 1–8.

Himmelsbach, M., 2012. 10 Years of MS instrumental developments—impact on LC-MS/MS in clinical chemistry. J. Chromatogr. B 883–884, 3–17.

Hjelmeland, A.K., Collins, T.S., Miles, J.L., Wylie, P.L., Mitchell, A.E., Ebeler, S.E., 2012. High throughput, sub ng/L analysis of haloanisoles in wines using HS-SPME with GC-triple quadrupole MS. Am. J. Enol. Vitic. 63, 494–499.

Hjelmeland, A.K., King, E.S., Ebeler, S.E., Heymann, H., 2013. Characterizing the chemical and sensory profiles of U.S. Cabernet Sauvignon wines and blends. Am. J. Enol. Vitic. 64, 169–179.

Hjelmeland, A.K., Wylie, P., Ebeler, S.E., 2016. A comparison of sorptive extraction techniques coupled to a new quantitative, sensitive, high throughput GC-MS/MS method for methoxypyrazine analysis in wine. Talanta 148, 336–345.

Hopfer, H., Ebeler, S.E., Heymann, H., 2012. How blending affects the sensory and chemical properties of red wine. Am. J. Enol. Vitic. 63 (3), 313–324.

Hopfer, H., Nelson, J., Collins, T.S., Heymann, H., Ebeler, S.E., 2015. The combined impact of vineyard origin and processing winery on the elemental profile of red wines. Food Chem. 172, 486–496.

Hopfer, H., Nelson, J., Mitchell, A.E., Heymann, H., Ebeler, S.E., 2013. Profiling the trace metal composition of wine as a function of storage temperature and packaging type. J. Anal. At. Spectrom. 28, 1288–1291.

Houk, R.S., 1986. Mass spectrometry of inductively coupled plasmas. Anal. Chem. 58, 97A–105A.

Hughey, C.A., McMinn, C.M., Phung, J., 2016. Beeromics: from quality control to identification of differentially expressed compounds in beer. Metabolomics 12, 11.

Intelmann, D., Haseleu, G., Hofmann, T., 2009. LC-MS/MS quantitation of hop-derived bitter compounds in beer using the ECHO technique. J. Agric. Food Chem. 57, 1172–1182.

Jennings, W.G., Poole, C.F., 2012. Milestones in the development of gas chromatography. In: Poole, C.F. (Ed.), Gas Chromatography. Elsevier, Waltham, MA.

Jennings, W.G., Rapp, A., 1983. Sample Preparation for Gas Chromatographic Analysis. Hüthig, Heidelberg.

Jennings, W.G., Shibamoto, T., 1980. Qualitative Analysis of Flavor and Fragrance Volatiles by Glass Capillary Gas Chromatography. Academic Press, New York.

Johnson, A., Hirson, G., Ebeler, S.E., 2012. Perceptual characterization and analysis of aroma mixtures using gas chromatography recomposition-olfactometry. PLoS One 7 (9), e42693.

Klee, M.S., 2012. Detectors. In: Poole, C.F. (Ed.), Gas Chromatography. Elsevier, Waltham, MA.

Knapp, D.R., 1979. Handbook of Analytical Derivatization Reactions. John Wiley and Sons, New York.

Lambert, M., Meudec, E., Verbaere, A., Mazerolles, G., Wirth, J., Masson, G., Cheynier, V., Sommerer, N., 2015. A high-throughput UHPLC-QqQ-MS method for polyphenol profiling in rosé wines. Molecules 20, 7890–7914.

Lea, A.G., Piggott, H.J., 2003. Fermented Beverage Production. Springer Science, New York.

Liger-Belair, G., Cilindre, C., Gougeon, R.D., Lucio, M., Gebefügi, I., Jeandet, P., Schmitt-Kopplin, P., 2009. Unraveling different chemical fingerprints between a champagne wine and its aerosols. PNAS 106, 16545–16549.

Luan, F., Hampel, D., Mosandl, A., Wüst, M., 2004. Enantioselective analysis of free and glycosidically bound monoterpene polyols in *Vitis vinifera* L. Cvs. Morio Muscat and Muscat Ottonel: Evidence for an oxidative monoterpene metabolism in grapes. J. Agric. Food Chem. 52, 2036–2041.

Luan, F., Mosandl, A., Gubesch, M., Wüst, M., 2006. Enantioselective analysis of monoterpenes in different grape varieties during berry ripening using stir bar sorptive extraction- and solid phase extraction-enantioselective-multidimensional gas chromatography-mass spectrometry. J. Chromatogr. A 1112, 369–374.

Mahmood, N., Petraco, N., He, Y., 2012. Elemental fingerprint profile of beer samples constructed using 14 elements determined by inductively coupled plasma-mass spectrometry (ICP-MS): multivariation analysis and potential application to forensic sample comparison. Anal. Bioanal. Chem. 402 (2), 861–869.

Majors, R.E., 2013. Sample Preparation Fundamental for Chromatography. Agilent Technologies, Wilmington, DE.

McDaniel, M.R., Miranda-Lopez, R., Watson, B.T., Michaels, N.J., Libbey, L.M., 1990. Pinot noir aroma: a sensory/gas chromatographic approach. In: Charalambous, G. (Ed.), Flavors and Off-Flavors. Elsevier Science, Amsterdam, pp. 23–26.

Miracle, R.E., Ebeler, S.E., Bamforth, C.W., 2005. The measurement of sulfur-containing aroma compounds in samples from production-scale brewery operations. J. Am. Soc. Brew. Chem. 63 (3), 129–134.

Molyneux, R.J., Schieberle, P., 2007. Compound identification: a *Journal of Agricultural and Food Chemistry* perspective. J. Agric. Food Chem. 55, 4625–4629.

Nelson, J., Hopfer, H., Gilleland, G., Boulton, R.B., Ebeler, S.E., 2015. Elemental profiling of Malbec wines under controlled winemaking conditions by microwave plasma atomic emission spectroscopy. Am. J. Enol. Vitic. 66 (3), 373–378.

NIST/NIH/EPA Mass Spectral Library, 2011. Standard Reference Database 1, NIST 11. Standard Reference Data Program, National Institute of Standards and Technology: Gaithersburg, MD, USA

NIST, 2016. NIST Chemistry WebBook. Standard Reference Database, National Institute of Science and Technology Standards. Available from: http://webbook.nist.gov/chemistry/name-ser.html

Nollet, L.M.L., Toldrá, F., 2013. Food Analysis by HPLC. CRC Press/Taylor & Francis Group, Boca Raton, FL.

Ochiai, N., Sasamoto, K., Kishimoto, T., 2015. Development of a method for the quantitation of three thiols in beer, hop, and wort samples by stir bar sorptive extraction with in situ derivatization and thermal desorption-gas chromatography-tandem mass spectrometry. J. Agric. Food Chem. 63, 6698–6706.

Oliver, J.D., Gaborieau, M., Castignolles, P., 2014. Ethanol determination using pressure mobilization and free solution capillary electrophoresis by photo-oxidation assisted ultraviolet detection. J. Chromatogr. A 1348, 150–157.

Panossian, A., Mamikonyan, G., Torosyan, M., Gabrielyan, E., Mkhitaryan, S., 2001. Analysis of aromatic aldehydes in brandy and wine by high-performance capillary electrophoresis. Anal. Chem. 73, 4379–4383.

Peng, Y., Chu, Q., Liu, F., Ye, J., 2004. Determination of phenolic constituents of biological interest in red wine by capillary electrophoresis with electrochemical detection. J. Agric. Food Chem. 52, 153–156.

Peris-Tortajada, M., 2000. HPLC determination of carbohydrates in foods. In: Nollet, L.M.L. (Ed.), Food Analysis by HPLC. Marcel Dekker, New York, NY, pp. 287–302.

Pollien, P., Fay, L.B., Baumgartner, M., Chaintreau, A., 1999. First attempt of odorant quantitation using gas chromatography-olfactometry. Anal. Chem. 71, 5391–5397.

Pollien, P., Ott, A., Montigon, F., Baumgartner, M., Muñoz-Box, R., Chaintreau, A., 1997. Hyphenated headspace-gas chromatography-sniffing technique: screening of impact odorants and quantitative aromagram comparisons. J. Agric. Food Chem. 45, 2630–2637.

Poole, C.F., 2012. Gas Chromatography. Waltham, MA, Elsevier.

Risticevic, S., Chen, Y., Kudlejova, L., Vatinno, R., Baltensperger, B., Stuff, J.R., Hein, D., Pawliszyn, J., 2010a. Protocol for the development of automated high-throughput SPME-GC methods for the analysis of volatile and semivolatile constituents in wine samples. Nat. Protoc. 5, 162e176.

Risticevic, S., Lord, H., Górecki, T., Arthur, C.L., Pawliszyn, J., 2010b. Protocol for solid-phase microextraction method development. Nat. Protoc. 5, 122e139.

Robinson, A.L., Adams, D.O., Boss, P.K., Heymann, H., Solomon, P.S., Trengove, R.D., 2011a. The relationship between sensory attributes and wine composition for Australian Cabernet Sauvignon wines. Aust. J. Grape Wine Res. 17, 327–340.

Robinson, A.L., Boss, P.K., Heymann, H., Solomon, P.S., Trengove, R.D., 2011b. Development of a sensitive non-targeted method for characterizing the wine volatile profile using headspace solid-phase microextraction comprehensive two-dimensional gas chromatography time-of-flight mass spectrometry. J. Chromatogr. A 1218, 504–517.

Robinson, A.L., Boss, P.K., Solomon, P.S., Trengove, R.D., Heymann, H., Ebeler, S.E., 2014. Origins of grape and wine aroma. Part 2. Chemical and sensory analysis. Am. J. Enol. Vitic. 65, 25–42.

Roullier-Gall, C., Boutegrabet, L., Gougeon, R.D., Schmitt-Kopplin, P., 2014a. A grape and wine chemodiversity comparison of different appelations in Burgundy: Vintage vs terroir effects. Food Chem. 152, 100–107.

Roullier-Gall, C., Lucio, M., Noret, L., Schmitt-Kopplin, P., Gougeon, R.D., 2014b. How subtle is the "Terroir" effect? Chemistry-related signatures of two "Climats de Bourgogne". PLoS One 9, e97615.

Russell, I., Stewart, G., 2014. Whisky: technology, second ed. Academic Press, New York.

Ryona, I., Pan, B.S., Sacks, G.L., 2009. Rapid measurement of 3-alkyl-2-methoxypyrazine content of winegrapes to predict levels in resultant wines. J. Agric. Food Chem. 57, 8250–8257.

Schieberle, P., Molyneux, R.J., 2012. Quantitation of sensory-active and bioactive constituents of food: *A Journal of Agricultural and Food Chemistry* perspective. J. Agric. Food Chem. 60, 2404–2408.

Schmarr, H.-G., Ganß, S., Koschinski, S., Fischer, U., Riehle, C., Kinnart, J., Potouridis, T., Kutyrev, M., 2010. Pitfalls encountered during quantitative determination of 3-alkyl-2-methoxypyrazines in grape must and wine using gas chromatography-mass spectrometry with stable isotope dilution analysis. Comprehensive two-dimensional gas chromatography-mass spectrometry and on-line liquid chromatography-multidimensional gas chromatography-mass spectrometry as potential loopholes. J. Chromatogr. A 1217 (43), 6769–6777.

Schoenmakers, P., Aarnoutse, P., 2014. Multi-dimensional separations of polymers. Anal. Chem. 86, 6172–6179.

Siebert, T.E., Smyth, H.E., Capone, D.L., Neuwöhner, C., Pardon, K.H., Skouroumounis, G.K., Herderich, M.J., Sefton, M.A., Pollnitz, A.P., 2005. Stable isotope dilution analysis of wine fermentation products by HS-SPME-GC-MS. Anal. Bioanal. Chem. 381, 937–947.

Skogerson, K., Runnebaum, R., Wohlgemuth, G., de Ropp, J., Heymann, H., Fiehn, O., 2009. Comparison of gas chromatography-coupled time-of-flight mass spectrometry and ^{1}H Nuclear Magnetic Resonance spectroscopy metabolite identification in white wines from a sensory study investigating wine body. J. Agric. Food Chem. 57, 6899–6907.

Skoog, D.A., Holler, F.J., Crouch, S.R., 2007. Principles of Instrumental Analysis, sixth ed. Thomson Brooks/Cole, Belmont, CA.

Stremel Azevedo, M., Pirassol, G., Fett, R., Micke, G.A., Vitali, L., Oliveira Costa, A.C., 2014. Screening and determination of aliphatic organic acids in commercial Brazilian sugarcane spirits employing a new method involving capillary electrophoresis and a semi-permanent adsorbed polymer coating. Food Res. Int. 60, 123–130.

Sumner, L.W., Amberg, A., Barrett, D., Beale, M.H., Beger, R., Daykin, C.A., Fan, T.W., Fiehn, O., Goodacre, R., Griffin, J.L., 2007. Proposed minimum reporting standards for chemical analysis. Metabolomics 3, 211–221.

Tswett, M., 1906a. Physikalisch-chemische Studien über das chlorophyll. Die adsorptionen. Ber Deutsch Botanisch Ges 24, 316–323.

Tswett, M., 1906b. Adsorptionanalyse und chromatographische method. Ansendung auf die Chemie des Chlorophylls. (Adsorption analysis and chromatographic method. Application to the chemistry of chlorophyll.) Berichte der Deutschen Botanischen Gesellschaft (Reports of the German Botanical Society) 24, 384–393.

Uliyanchenko, K., Cools, P.J.C.H., van der Wal, S., Schoenmakers, P.J., 2012. Comprehensive two-dimensional ultrahigh-pressure liquid chromatography for separations of polymers. Anal. Chem. 84, 7802–7809.

Ullrich, F., Grosch, W., 1987. Identification of the most intense volatile flavor compounds formed during autoxidation of linoleic acid. Z. Lebensm. Unters. Forsch 184, 277–282.

Wang, M.-L., Choong, Y.-M., Su, N.-W., Lee, M.-H., 2003. A rapid method for determination of ethanol in alcoholic beverages using capillary gas chromatography. J. Food Drug Anal. 11, 133–140.

Webb, K., Bristow, T., Sargent, M., Stein, B., 2004. Methodology for Accurate Mass Measurement of Small Molecules. Best Practices Guide, LGC Limited, Teddington, UK. Available from: http://www.bmss.org.uk/Docs/VIMMS_guide.pdf

Willemse, C.M., Stander, M.A., Vestner, J., Tredoux, A.G.J., de Villiers, A., 2015. Comprehensive two-dimensional hydrophilic interaction chromatography (HILIC) x reversed-phase liquid chromatography coupled to high-resolution mass spectrometry (RP-LC-UV-MS) analysis of anthocyanins and derived pigments in red wine. Anal. Chem. 87, 12006–12015.

Wisniewska, P., Sliwinska, M., Dymerski, T., Wardencki, W., Namiesnik, J., 2016. Qualitative characteristics and comparison of volatile fraction of vodkas made from different botanical materials by comprehensive two-dimensional gas chromatography and the electronic nose based on the technology of ultra-fast gas chromatography. J. Sci. Food Agric., [Epub ahead of print].

Wu, Q., Yuan, H., Zhang, L., Zhang, Y., 2012. Recent advances on multidimensional liquid chromatography-mass spectrometry for proteomics: from qualitative to quantitative analysis—a review. Anal. Chim. Acta 731, 1–10.

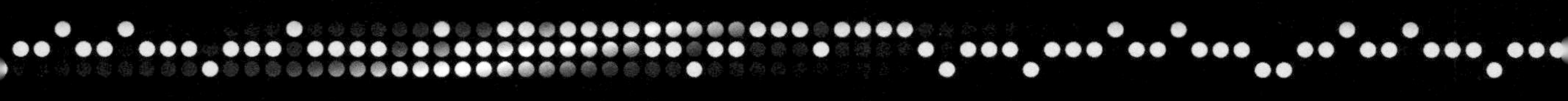

In this chapter

Objective

To determine if trained panelists could discriminate a 1% (v/v) alcohol difference in a Chardonnay wine.

Case Study 1

The effect of alcohol reduction on an oaked Chardonnay wine: discrimination[a]

1.1 Materials and methods

A group of panelists were trained to do a sensory descriptive analysis (DA) of 11 Chardonnay wines that had been produced by blending a parent wine with a de-alcoholized version of the parent in approximately 0.2% (v/v) increments. Eleven of these panelists were then asked to participate in a series of discrimination tests. Each panelist performed four replicates of each of the three 2-alternative forced choice (2AFC) tests, in each case they were asked which sample was higher in perceived alcohol content (hot mouthfeel). As described in the Sensory Methodology Chapter 3, using 2AFC is extremely rare with real food since these foods usually change in multiple ways and this makes the 2AFC not an appropriate test choice. However, in this case we had completed the DA and knew that the wines only differed significantly in overall aroma intensity and in hot mouthfeel. Therefore we were comfortable using the 2AFC test. In test 1 they compared a 12.93 ± 0.02% (v/v) wine (W1) with the parent wine at 14.94 ± 0.01% (v/v) (W11). In test 2 they compared W1 with a wine at 13.99± 0.01% (v/v) (W6) and in test 3 W11 wine was compared to W6. Thus in test 1 the alcohol difference was about 2% and in tests 2 and 3 it was about 1%. During each of the four sessions the panelists received the three tests in a randomized order and the samples within a test were also randomized.

The discrimination tests were conducted in isolated and ventilated tasting booths, under white light, in the J. Lohr wine sensory laboratory of the University of California, Davis. Each sample was presented at a constant volume (25 mL) at room temperature, in clear, covered, ISO 3591:1977 tasting glasses labeled with random three digit codes that differed for each panelist. Assessors were asked

[a]Based on data used in King and Heymann (2014).

Sensory and Instrumental Evaluation of Alcoholic Beverages. http://dx.doi.org/10.1016/B978-0-12-802727-1.00007-7

to expectorate all samples. FIZZ software (Version 2.1, Biosystèmes, Couternon, France) was used for the collection of all data.

1.2 Data analysis

Table CS1.1 shows the raw data used in this case study. For the King and Heymann (2014) publication the aggregate data were analyzed and we do so here using the *z*-test described in the Sensory Methodology chapter.

$$z = \frac{(X - np - 0.5)}{\sqrt{npq}} \tag{CS1.1}$$

where X = number of correct responses; N = total number of responses; p = probability of correct decision by chance and in the case $p = 1/2$; $q = (1 - p)$ and 0.05 is a continuity correction. All three of the 2AFC tests had 29 out of 44 correct responses.

$$z = \frac{(29 - 44(0.5) - 0.5)}{\sqrt{44(0.5)(0.5)}} = \frac{6.5}{\sqrt{11}} = 1.9598 = 1.96 \tag{CS1.2}$$

Table CS1.1 Raw data from the 2AFC test

Rep	Panelist	W1 and W11	W1 and W6	W6 and W11
1	RD	1[a]	1	1
1	WD	0	0	1
1	MJ	0	0	0
1	ZZ	1	0	1
1	HD	1	1	1
1	HA	1	1	0
1	AM	1	1	1
1	DM	0	1	1
1	WP	1	1	1
1	WK	1	0	0
1	CC	1	1	0
2	RD	0	1	0
2	WD	0	1	1
2	MJ	1	0	1
2	ZZ	0	0	1
2	HD	0	1	0

Rep	Panelist	W1 and W11	W1 and W6	W6 and W11
2	HA	0	0	1
2	AM	1	0	1
2	DM	1	0	1
2	WP	1	0	1
2	WK	0	1	1
2	CC	1	1	0
3	RD	1	0	0
3	WD	1	1	1
3	MJ	0	1	1
3	ZZ	1	1	1
3	HD	1	0	1
3	HA	1	1	0
3	AM	1	1	1
3	DM	1	0	1
3	WP	1	1	1
3	WK	1	1	0
3	CC	1	0	1
4	RD	0	1	1
4	WD	0	1	0
4	MJ	0	0	0
4	ZZ	1	1	1
4	HD	1	1	1
4	HA	0	1	1
4	AM	0	1	0
4	DM	1	1	0
4	WP	1	1	1
4	WK	1	1	0
4	CC	1	1	1

[a] *1 indicates panelist correctly identified the higher alcohol wine.*
W1 had 12.93% ABV, W6 had 13.99% ABV, and W11 had 14.94% ABV. ABV, alcohol by volume; 2AFC, 2-alternative forced choice.

From the z-table the cumulative probability for this value is 0.9750. Thus the alpha probability of this outcome is 1 − 0.9750 or 0.025. Therefore all three of these tests were significant and the trained panelists could tell the perceptual difference between samples that differed in about 1% (v/v) alcohol. We could also use sensR (https://cran.r-project.org/web/packages/sensR/index.html) (based on the paper by Brockhoff and Christensen, 2010) to analyze the aggregate data using the following code:

```
# Discrimination Testing Data Analysis
# All code comes without any warranty
# We had 11 panelists evaluating each 2AFC sample pair 4 times. For the
# aggregate data the higher alcohol sample was correctly identified 29
# times.
# We are using the sensR package

library(sensR)

#The discrim function computes the probability of a correct answer (Pc),
#the probability of discrimination (Pd) and d-prime, their standard
#errors,confidence intervals and a p-value of a difference for a 2AFC
#("twoAFC"). #This function can also calculate the same set of values for
#"duo-trio", #"triangle", "tetrad", and "threeAFC" discrimination tests.
m=discrim(29, 44, method ="twoAFC")
print(m)
```

This leads to the following output:

```
Estimates for the twoAFC discrimination protocol with 29 correct
answers in 44 trials. One-sided p-value and 95 % two-sided confidence
intervals are based on the 'exact' binomial test.

        Estimate Std. Error    Lower  Upper
pc        0.6591    0.07146 0.500805 0.7951
pd        0.3182    0.14292 0.001610 0.5902
d-prime   0.5798    0.27553 0.002854 1.1656

Result of difference test:
'exact' binomial test:  p-value = 0.02438
Alternative hypothesis: d-prime is greater than 0
```

The "exact" binomial test (which is a one-tailed test for the 2AFC) gives us an alpha level of 0.02438 which is very close to value we calculated using the z-test.

1.3 Conclusions

In the aggregate the 11 panelists doing 4 replications could significantly identify the higher alcohol wine in all cases. Therefore, trained panelists could discriminate between the same wines with a reduced alcohol content of about 1% (v/v).

As an aside: The sensR program is also useful early in the experimental design since the program contains a function (discrimSS) that would calculate the number of observations needed for specified levels of alpha, beta, and proportion of discriminators (or the d-prime). Another function (discrimPwr) allows one to calculate the power (probability of detecting a difference if one existed) given the proportion of discriminators (or the d-prime), the alpha level and the number of observations. Using the discrimSS function we calculated the number of panelists needed based on eight different scenarios.

```
discrimSS(0.1, pd0 = 0, target.power = 0.90, alpha = 0.05,
          pGuess = 1/2, test = "difference", statistic = "exact")
discrimSS(0.3, pd0 = 0, target.power = 0.90, alpha = 0.05,
          pGuess = 1/2, test = "difference", statistic = "exact")
discrimSS(0.1, pd0 = 0, target.power = 0.80, alpha = 0.05,
          pGuess = 1/2, test = "difference", statistic = "exact")
discrimSS(0.3, pd0 = 0, target.power = 0.80, alpha = 0.05,
          pGuess = 1/2, test = "difference", statistic = "exact")
discrimSS(0.1, pd0 = 0, target.power = 0.90, alpha = 0.05,
          pGuess = 1/3, test = "difference", statistic = "exact")
discrimSS(0.3, pd0 = 0, target.power = 0.90, alpha = 0.05,
          pGuess = 1/3, test = "difference", statistic = "exact")
discrimSS(0.1, pd0 = 0, target.power = 0.80, alpha = 0.05,
          pGuess = 1/3, test = "difference", statistic = "exact")
discrimSS(0.3, pd0 = 0, target.power = 0.80, alpha = 0.05,
          pGuess = 1/3, test = "difference", statistic = "exact")
```

From Table CS1.2 it is clear that requiring both a high level of power and a very low proportion of discriminators in the population leads to the need for

Table CS1.2 A priori calculations indicating the number of observations needed for differing values of proportion of discriminators, power, and alpha levels

Proportion of discriminators (%)	Required power (%)	Required alpha	Guessing proportion	Number of observations needed
10	90	0.05	0.5 (duo–trio)	866
30	90	0.05	0.5 (duo–trio)	93
10	80	0.05	0.5 (duo–trio)	620
30	80	0.05	0.5 (duo–trio)	69
10	90	0.05	0.33 (triangle)	447
30	90	0.05	0.33 (triangle)	53
10	80	0.05	0.33 (triangle)	325
30	80	0.05	0.33 (triangle)	40

very high numbers of observations. Increasing the proportion of discriminators or decreasing the power dramatically decreases the number of observations needed. Therefore, prior to doing a discrimination study the sensory scientist should carefully decide which of these values are required for the specific experiment. When the potential samples are very subtly different then the proportion of discriminators would by definition be very small; on the other hand samples that are quite different may naturally have a larger proportion of discriminators. Additionally, the level of power would allow the experimenter to be confident that if the outcome of the study is not to reject the null hypothesis that there was enough power to find a difference if one existed.

We also calculated the power for a series of studies with 50 observations, either 10 or 30% proportion of discriminators and using either duo–trio (or 2AFC) or triangle tests.

```
discrimPwr(0.1, pd0 = 0, 50, alpha = 0.05, pGuess = 1/2,
           test = "difference", statistic = "exact")
discrimPwr(0.3, pd0 = 0, 50, alpha = 0.05, pGuess = 1/2,
           test = "difference", statistic = "exact")
discrimPwr(0.1, pd0 = 0, 50, alpha = 0.05, pGuess = 1/3,
           test = "difference", statistic = "exact")
discrimPwr(0.3, pd0 = 0, 50, alpha = 0.05, pGuess = 1/3,
           test = "difference", statistic = "exact")
```

Table CS1.3 shows the output for the aforementioned code. The discrimPwr function would only be used if the outcome of the study had been not to reject the null hypothesis that there is no perceptual difference between the samples. If there were subtle differences for the duo-trio scenario the power would have been about 13%. Thus the experimenter had about a 13% chance to find a perceptual difference *if one existed.*

Table CS1.3 Posthoc power calculations for specified proportion of number of observations, discriminators, and alpha levels

Proportion of discriminators (%)	Number of observations	Required alpha	Guessing proportion	Calculated power
10	50	0.05	0.5 (duo–trio)	0.1273451
30	50	0.05	0.5 (duo–trio)	0.6215871
10	50	0.05	0.33 (triangle)	0.233983
30	50	0.05	0.33 (triangle)	0.8811288

References

Brockhoff, P.B., Christensen, R.H.B., 2010. Thurstonian models for sensory discrimination tests as generalized linear models. Food Qual. Prefer. 21, 330–338.

King, E.S., Heymann, H., 2014. The effect of reduced alcohol on the sensory profiles and consumer preferences of white wine. J. Sens. Stud. 29, 33–42.

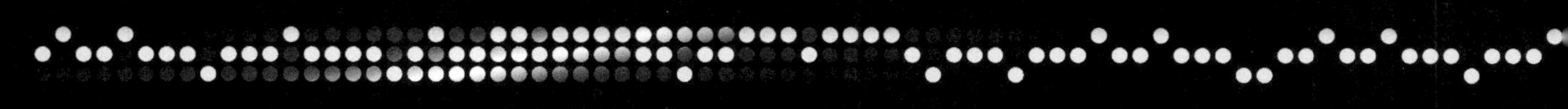

In this chapter

Objective

To determine whether inexpensive (less than AUS$10 per 750 mL bottle) differed significantly in their sensory attributes.

Case Study 2

A descriptive analysis[a] of inexpensive Australian Chardonnay wines[b]

2.1 Materials and methods

A subset of seven commercially available Australian Chardonnay wines, all less than AUS$10 per 750 mL bottle, were chosen for this case study (Table CS2.1) (In Table CS2.1 and in all data output the original numbering for the complete set of 21 wines have been retained. This was done to make it easier for anyone interested in comparing the results of the subset to the entire set of wines.). The 12-member sensory descriptive panel (6 men, 6 women ranging in age from 21 to 50 years) was selected based on interest and availability. Most of the panelists had extensive descriptive analysis experience with Australian Semillon wines but not with Chardonnay. Over six 1-h training sessions the panel smelled and tasted all the wines in the study and through consensus created the descriptors and the reference standards used (Table CS2.2).

For evaluation, the wines were served in blocks of seven. All wines were evaluated in triplicate. Wines were served to panelists in a completely randomized order within a block. Wines were coded with three digit random numbers. Wines were served in 25 mL aliquots in ISO wine glasses covered with glass Petri dish lids. The wines were served at ambient temperature (20°C) in booths equipped with red lights and a computerized data collection system (Compusense, Guelph, Ontario, Canada). All wines were expectorated, and water and crackers were provided as palate cleansers.

---

[a] *Since this is the first case study involving descriptive analysis (DA), we will show all MANOVA and ANOVA results tables, as well as the CVA plots. We will carefully interpret and explain the data. In later case studies involving DA data we will be more parsimonious and less detailed.*
[b] *Based on data used in Saliba et al. (2013).*

Sensory and Instrumental Evaluation of Alcoholic Beverages. http://dx.doi.org/10.1016/B978-0-12-802727-1.00008-9

Table CS2.1 Wines used for the Chardonnay sensory descriptive analysis

Wines[a]	
W1	Victoria 2009 ($22)
W2	Orange 2009 ($12; cc)
W3	Yarra Valley 2010 ($26)
W4	**Griffith 2010 ($7)**
W5	South Eastern Australia 2010 ($10)
W6	South Australia 2010 ($11, unwooded)
W7	**South Australia 2010 ($9)**
W8	Adelaide Hills 2010 ($16; cc)
W9	**South Eastern Australia 2010 ($9)**
W10	Orange 2010 ($16; cc)
W11	Western Australia 2010 ($14, unwooded)
W12	**South Eastern Australia 2011 ($4)**
W13	South Australia 2009 ($13)
W14	South Eastern Australia 2010 ($10)
W15	**South Eastern Australia 2010 ($9)**
W16	Adelaide Hills 2009 ($16; cc)
W17	**South Eastern Australia 2010 ($9)**
W18	**Riverland 2010 ($8)**
W19	Tumbarumba 2009 ($17; cc)
W20	Hunter Valley 2006 ($48)
W21	Riverland 2009 ($10)

[a]*All prices are rounded to the nearest AUS$; cc, cool climate. Wines chosen for the case study are indicated in* **bold**. *The original numbering for the complete set of 21 wines has been retained. This was done to make it easier for anyone interested in comparing the results of the subset to the entire set of wines (Saliba et al., 2013).*

2.2 Data analyses

Our standard data analysis strategy for sensory descriptive data is (1) establishing overall differences among wines using multivariate analysis of variance (MANOVA), (2) determining individual attributes that differ significantly with univariate analysis of variance (ANOVAs), and (3) establishing multivariate sample differentiation with canonical variate analysis (CVA) or principal component analysis (PCA) (We will describe the use of the PCA in Case Study 3 and then also discuss when we prefer to use one over the other.).

Table CS2.2 Attributes and reference standards used to train the Chardonnay sensory descriptive panel

Attribute	Reference standard
Confectionary	2 Allen's Red Frogs + 1 Musk Stick (broken into 8 pieces), NO WINE
Green apple	12.5 g Granny Smith Apple wedge; cut into 4 pieces + 25 mL wine*
Lemon/lime	1″ × 1″ lemon peel + 1″ × 1″ lime peel + 50 mL wine
Grapefruit	25 mL Original Juices Chilled Grapefruit Juice, NO WINE
Peach	25 g SPC Peach Halves in fruit juice, cut into 6 pieces + 25 mL wine
Apricot	25 g SPC Apricot Halves in fruit juice, cut into 6 pieces + 25 mL wine
Tropical fruit	25 mL syrup Dole Chunks of Tropical Fruit in light syrup, NO WINE
Melon	10 g Honeydew Melon, cut into 4 pieces, NO WINE
Fresh veg.	10 g McCains Frozen Sliced Green Beans + 10 g McCains Frozen Green Peas, NO WINE
Honey	10g Woolworths Pure Blend Honey + 25 mL hot water, NO WINE
Butter	14 g Woolworths Tasmanian Butter + 25 mL wine
Vanilla	2 mL Queen Imitation Vanilla + 25 mL wine
Straw	2 g chaff, NO WINE
Oak	10 g single toast oak chips (Double Chauffe, Boise France; Maumusson Laguian) soaked in 200 mL wine overnight; Decanted; Pour 25 mL
Char	45 g double toasted oak chips (Single Chauffe, Boise France; Maumusson Laguian) soaked in 200 mL wine overnight; Decanted; Pour 25 mL
Sour	1.0 g tartaric acid in 400 mL wine, pour 25 mL per panelist
Bitter	0.1 g quinine in 400 mL wine, pour 25 mL per panelist
Sweet	30 g sucrose in 400 mL wine, pour 25 mL per panelist
Astringent	0.02 g tannin in 600 mL wine, pour 25 mL per panelist
Creamy mouthfeel	Defined as a smooth, velvety, creamy feeling in the mouth

*Base wine: Sunnyvale Dry White Wine 4 L Bag in the Box. Miranda Wines, Merbein, VIC.

The data were analyzed by three-way MANOVA with two-way interaction (main effects: Panelist P, Wine W, and Replication R). The MANOVA wine main effect was found to be significant, therefore each attribute was subsequently analyzed by three-way ANOVA with all two-way interactions (main effects: Panelist P, Wine W, and Replication R; interaction effects: P × W, P × R, W × R). Any attribute with a significant wine main effect and a significant wine interaction were further evaluated with a pseudomixed model (Naes and Langsrud, 1998) using the mean square value of the significant interaction effect as the denominator. If the wine main effect was still significant then that attribute was deemed to be significant. Fisher's protected least significant difference (lsd) values were calculated for each significant wine attribute.

A one-way MANOVA (main effect wine) was then calculated and submitted to a CVA. Monrozier and Danzart (2001) have shown that simple one-way MANOVA models lead to more robust canonical variate analyses than more complex models. All analyses were performed in R and R-Studio.

2.3 Results

The three-way MANOVA with two-way interactions for the entire data set was significant at $p < 0.05$ for all sources of variation except the wine by replication interaction (Table CS2.3), thus there are significant sensory differences among the wines. Since we are interested in the main factor wine we can now go ahead and calculate the three-way ANOVAs with two-way interaction for all attributes. The individual attribute ANOVA tables are typically not published but to demonstrate the analyses we are showing the complete ANOVA for the attribute "confectionary" in Table CS2.4. This attribute had significant effects for the wine, judge, and wine by judge interaction terms. In most sensory panels we find (and actually expect) a significant panelists effect since, as seen Chapter 1, panelists will differ in their sensory systems. We are not concerned when we see these significant effects. Wine was also a significant source of variation and since we are interested in how the wines differ this is good news. However, the wine by judge interaction is disconcerting since it indicates that at least one panelist evaluated

Table CS2.3 Three-way multivariate analysis of variance (MANOVA) with all two-way interactions

Factor	df[a]	Wilks lambda	F[b]	Num[c] df	Den[d] df	Probabilty (>F)	Significance
Wine	6	0.11365	2.5147	120	660.53	1.57E-13	*[e]
Rep	2	0.57307	1.8136	40	226	0.003745	*
Judge	11	0.00001	11.4016	220	1120.94	2.20E-16	*
Wine:rep	12	0.17658	0.9124	240	1198.99	0.811865	
Wine:Judge	66	0.00002	1.2967	1320	2299.76	3.73E-08	*
Rep:judge	22	0.01473	1.3055	440	1747.39	0.000136	*
Residuals	132						

[a] *Study degrees of freedom.*
[b] *Approximate* F-*value.*
[c] *Numerator degrees of freedom.*
[d] *Denominator degrees of freedom.*
[e] *Factor significant at alpha = 0.05.*

Table CS2.4 Three-way analysis of variance (ANOVA) with all two-way interactions for confectionary

Factor	df	Sum of squares	Mean squares	*F*-value	Probabilty (>*F*)	Significance
Wine	6	126.86	21.1426	7.649	4.63E-07	* (*)
Rep	2	4.32	2.1586	0.7809	0.46008	
Judge	11	294.05	26.7318	9.6711	1.17E-12	*
Wine:rep	12	38.9	3.2413	1.1726	0.30916	
Wine:judge	66	283.73	4.2989	1.5553	0.01644	*
Rep:judge	22	75.28	3.4218	1.2379	0.22761	
Residuals	132	364.86	2.7641			

* (*) indicates that this attribute remained significant after testing with the pseudomixed model.

at least one wine differently from the rest of the panelists. This type of interaction could be very important and may make the significant wine effect irrelevant. To test the importance of the significant wine by judge interaction term we calculate a new *F*-value using a pseudomixed model (Naes and Langsrud, 1998). Specifically an *F*-value for the wine main effect is calculated using the mean square of the significant interaction as the denominator. If the new *F*-value is still significant at $p < 0.05$ then the interaction effect is not important and we will continue to say that the wine effect is significant. On the other hand if the new *F*-value is now not significant then the interaction effect is very important and the wine effect is not significant anymore.

When we calculate the pseudomixed model for the attribute confectionary:

$$F\text{-value} = 21.1426 / 4.2989 = 4.918$$

Critical *F*-value (degrees of freedom numerator 6, degrees of freedom denominator 66) = 2.24.

For the attribute confectionary the pseudomixed model indicated that the wine effect was *still* significant, thus we conclude that there are significant differences in this attribute among the tested wines.

Table CS2.5 shows a consolidated ANOVA table for all attributes as well as the effect of the pseudomixed models when there were significant wine by judge and/or wine by replication interactions. The attributes that were significant sources of variation among the different wines were confectionary, green apple, peach, apricot, tropical fruit, melon, fresh veg., honey, butter, oak, bitter, and sour.

Fisher's protected lsd values were then calculated for these significant attributes and the means and the mean comparisons based on these values are shown in Table CS2.6. W12 had the most intense confectionary, green apple, peach,

Table CS2.5 Summary of significance at $p < 0.05$ for three-way ANOVAs with two-way interactions for all attributes

Fcator	df	Confectionary	Green apple	Lemon lime	Grape fruit	Peach	Apricot	Tropical fruit
Wine	6	* (*)	*			*	*	*
Rep	2							*
Judge	11	*	*	*	*	*	*	*
Wine:rep	12							
Wine:judge	66	*	*					
Rep:judge	22							*
Residuals	132							

Factor	df	Melon	Fresh veg.	Honey	Butter	Vanilla	Straw	Oak
Wine	6	*	*	* (*)	* (*)	* (*)	* (NS)	*
Rep	2	*					*	
Judge	11	*	*	*	*	*	*	*
Wine:rep	12			*				
Wine:judge	66			*	*	*	*	
Rep:judge	22						*	*
Residuals	132							

Factor	df	Char	Sour	Bitter	Sweet	Astringent	Creamy mouthfeel
Wine	6	* (NS)		*	* (*)		
Rep	2				*		*
Judge	11	*	*	*	*	*	*
Wine:rep	12						
Wine:judge	66	*			*	*	*
Rep:judge	22	*	*			*	
Residuals	132						

* (*) indicates that this attribute remained significant after testing with the pseudomixed model; * (NS) indicates that this attribute was not significant after testing with the pseudomixed model.

Table CS2.6 Means for all significant attributes with Fisher's protected least significant difference (lsd)

Wine	Confectionary		Green apple		Peach		Apricot		Tropical fruit	
W04	3.3	b	2.5	abc	3.2	b	2.5	ab	3	b
W07	2.7	bc	2.7	ab	2.4	cd	2.2	b	2.2	c
W09	2.0	c	2.1	bcd	2.2	cd	2	b	2.1	c
W12	4.1	a	2.8	a	3.9	a	3.1	a	4.4	a
W15	2.0	c	1.7	d	1.8	d	1.9	b	1.9	c
W17	2.4	c	2.1	cd	2.6	bc	2.2	b	2.2	c
W18	2.7	bc	2	cd	2.5	bc	2.2	b	2.2	c
lsd	0.77		0.58		0.7		0.67		0.67	

Wine	Melon		Fresh veg.		Honey		Butter		Vanilla	
W04	2.1	b	0.7	cd	1.5	ab	1.8	ab	2.3	ab
W07	1.7	bc	1.2	ab	1.5	b	1.3	cd	1.8	bc
W09	2.1	bc	0.8	bcd	1.6	ab	2.1	ab	2.7	a
W12	2.9	a	1.3	a	1.4	bc	1.1	d	1.4	c
W15	1.5	c	0.8	cd	2	a	1.7	bc	1.9	bc
W17	1.7	bc	0.5	d	1	c	1.7	abc	1.9	b
W18	1.7	bc	1	abc	1.1	bc	2.2	a	2.1	b
lsd	0.6		0.47		0.49		0.48		0.56	

Wine	Oak		Bitter		Sweet	
W04	2.2	b	2.4	bc	3.3	b
W07	2.1	b	3	ab	2.9	bc
W09	2.9	a	2.6	b	2.9	bc
W12	1.6	c	1.9	c	4	a
W15	2.3	ab	2.9	ab	2.4	c
W17	2.3	b	3.3	a	2.7	bc
W18	2.6	ab	2.7	ab	2.7	bc
lsd	0.55		0.67		0.57	

Wines with the same letters within an attribute are not significantly different at $p < 0.05$.

apricot, tropical fruit, melon, and fresh veg. aromas and it had the sweetest taste. This wine was the youngest wine in the product set and had been produced and bottled approximately 2 months prior to the DA study.

We usually prefer to evaluate the results of the means table (Table CS2.6) in conjunction with a multivariate plot created by a CVA or by a PCA. Prior to performing the CVA we ran a one-way MANOVA with wine as the only factor and the Wilks lambda value for wine was 0.44147 with an *F*-value of 2.6453 which, given the degrees of freedom of 78 and 1290.8, was significant at $p < 0.05$. We then performed the CVA and the output is shown in Table CS2.7. Usually this output would not be published and the authors would just refer to it; since we are showing as much information as possible in this case study we are including it here.

The first canonical variate (CV1) explained 69.2% of the variance ratio, while CV2 explained an additional 11.9% with CV3 adding another 10%. Unlike PCA where there is no test to determine whether one or more principal components (PCs) are significant in describing the multivariate space, CVA has Bartlett's test (Bartlett, 1938). Bartlett's test is based on the Wilks lambda value associated with the MANOVA and allows one to determine how many (or if any) CVs are

Table CS2.7 Output of the canonical variate analysis (CVA) showing the squared canonical correlations, the eigenvalues, the percent variance ratio explained, Bartlett's test statistics, the approximate *F*-values, numerator and denominator degrees of freedom, and probability for each CV

CV	CanRsq	Eigenvalue	Difference	Percent	Cumulative
1	0.407818	0.688671	0.57053	69.2215	69.221
2	0.10566	0.118143	0.57053	11.8751	81.097
3	0.089894	0.098773	0.57053	9.9282	91.025
4	0.038865	0.040436	0.57053	4.0645	95.089
5	0.032511	0.033604	0.57053	3.3777	98.467
6	0.015024	0.015253	0.57053	1.5331	100
CV	**Test stat.**	**Approx *F***	**Num. df**	**Den. df**	**Probability**
1	0.44147	6.0071	36	1056.67	<2.2e-16
2	0.74551	2.9511	25	896.78	2.07E-06
3	0.83358	2.8392	16	739.96	0.000165
4	0.91592	2.4153	9	591.55	0.010705
5	0.95295	2.9752	4	488	0.019047
6	0.98498	3.7369	1	245	0.054375

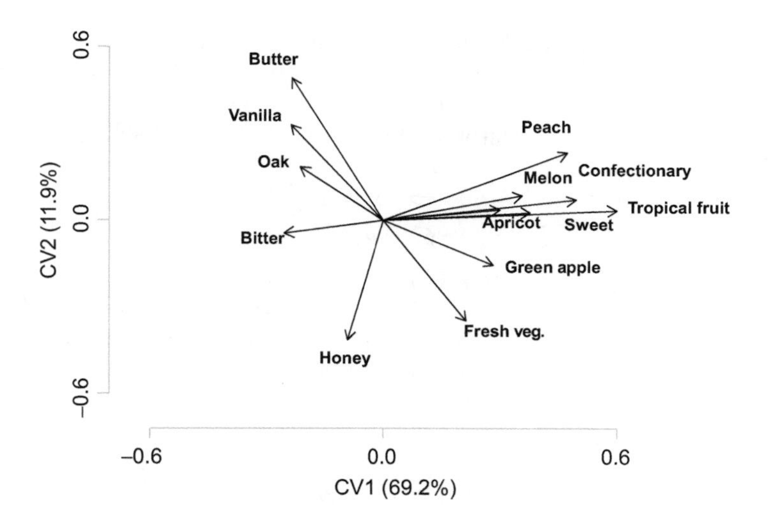

Figure CS2.1
CVA loadings plot of Chardonnay wines
CV1 versus CV2.

significant. In this instance five of the six canonical variates were significant but we chose to only plot the first three since they combined to explain over 90% of the variance ratio. Figs. CS2.1 and CS2.2 show the loadings and scores plots, respectively for the first two dimensions of the CVA. In the loadings plot the distribution of the sensory attributes are shown in the CVA space. Attribute vectors that differ from each other with a small positive angle are highly positively correlated (e.g., apricot and confectionary, tropical and sweet, melon and confectionary, or green apple and fresh veg.) and those with small negative angles are highly negatively correlated (e.g., bitter and sweet, bitter and melon, butter and fresh veg., or oak and green apple). Lastly, attribute vectors at a 90-degree angle are not correlated (e.g., peach and butter, peach and fresh veg.). Longer attribute

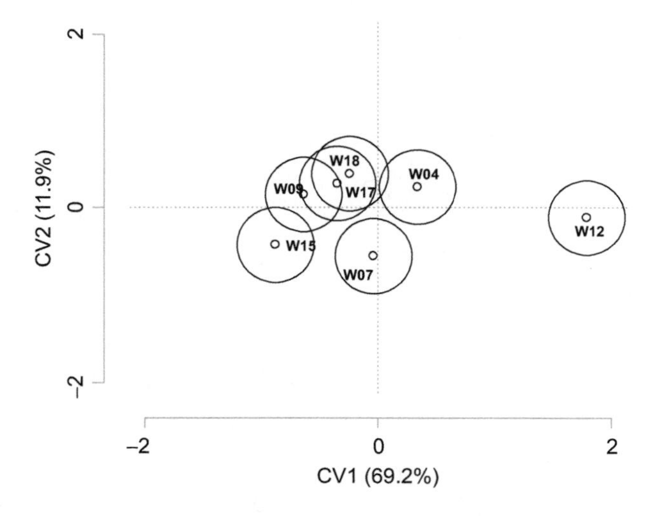

Figure CS2.2
CVA scores plot of Chardonnay wines
CV1 versus CV2. See Table CS2.1 for wine codes.

vectors are more important in defining the space than shorter ones. Based on this subset of the data the attributes melon, confectionary, tropical fruit, and apricot seem to be redundant that is, are describing similar sensory properties, and one of these aroma terms would have been sufficient to describe the positive side of CV1. Please note that sweet (a taste attribute) is not included in these redundant aroma attributes. Thus we would say that CV1 is a fruity/sweet versus bitter axis and the CV2 axis is defined by honey/fresh veg. opposed to butter aromas.

The scores plot (Fig. CS2.2) shows the locations of the products (i.e., the wine samples) in the CVA space. We have also included the 95% confidence intervals (CIs) for each product. On CV1 and CV2, W12, a very fruity and sweet wine (to double check our interpretation of the CVA space we always need to check the univariate mean values in Table CS2.6) with low bitterness, is separated from all other wines. W7, a wine balanced in nearly all attributes, since it is close to the fulcrum (or plot origin) for CV1 and CV2 and it is also separated from all other wines. The CIs for W15 and W9 intersect and these two wines are higher in bitter taste and lower in the fruity/sweet sensory attributes. The CIs for W4 and W18 also intersect and these two wines are quite high in Vanilla. Additionally, the CIs for W9, W17, and W18 also intersect and based solely on the first two dimensions of the CVA seem to have similar sensory attributes. However, based on the means (Table CS2.6) the most bitter wine was W17—something not shown by the first two dimensions of the CVA. Therefore, since CV3 explained nearly as much of the variance ratio as CV2 (10% vs. 11.2%) we plotted CV1 versus CV3 (Figs. CS2.3 and CS2.4) and based on these plots W17 is more closely correlated to the bitter loadings vector. Additionally, it is clear that the CI for W9 does not intersect with either of the CIs of W17 and W18, that is, W9 and W5 are above the plane and W18 and W17 are below. The CV3 axis is likely a honey/vanilla versus bitter axis. Additionally, based on Fig. CS2.3 the redundant terms are

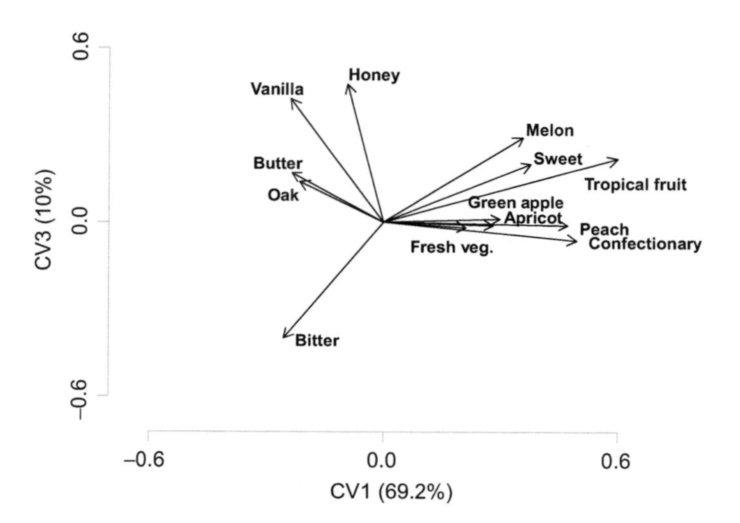

Figure CS2.3
CVA loadings plot of Chardonnay wines
CV1 versus CV3.

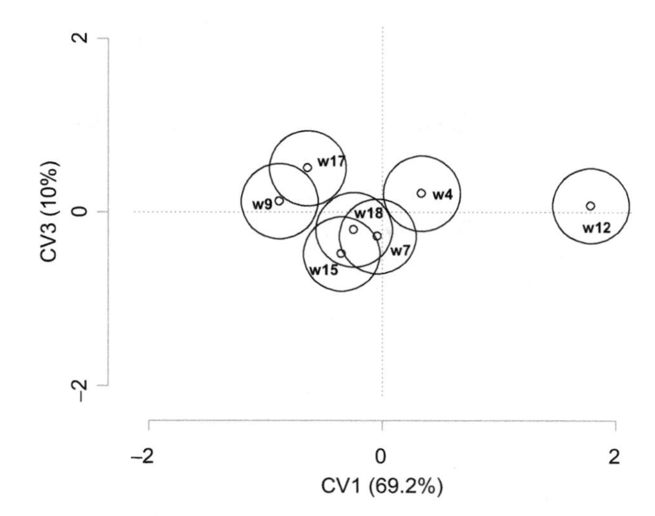

Figure CS2.4
CVA scores plot of Chardonnay wines
CV1 versus CV3. See Table CS2.1 for wine codes.

green apple, peach, apricot, and confectionary but not tropical fruit and peach. This is why we always look at plots one level above the ones that we will use in the publication to make sure that we are seeing all the differences among the samples. Thus based on the first three CVs we have five groups of wine in the multidimensional space: W12, W7, W15 and W9, and W4 and W18 with W18 also intersecting with W17. However, based on the means the major differences were W12 versus the rest of the Chardonnay wines.

2.4 Conclusions

The inexpensive Chardonnay wines tested in this study differed in their sensory attributes. In general, the wines were distinguished by fruity/sweet attributes, bitter taste, and butter, honey/fresh veg. aromas. Based on the first three CVs we have five groupings of wine in the multidimensional space, with sensory attributes differing among the grouping: W12; W7; W15 and W9; and W4, W18, and W17. However, based on the sample means the major differences were W12 versus the rest of the Chardonnay wines. W12 was the most fruity/sweet of the tested wines.

References

Bartlett, M.S., 1938. Further aspects o f the theory of multiple regression. Proc. Cambridge Philos. Soc. 34, 33–40.

Monrozier, R., Danzart, M., 2001. A quality measurement for sensory profile analysis—the contribution of extended cross-validation and resampling techniques. Food Qual. Prefer. 12, 393–406.

Naes, T., Langsrud, Ø., 1998. Fixed or random assessors in sensory profiling? Food Qual. Prefer. 9, 145–152.

Saliba, A.J., Heymann, H., Blackman, J.W., MacDonald, J.W., 2013. Consumer-sensory evaluation of Australian Chardonnay. Wine Vitic. J. May/June, 64–66.

```
# all code comes without any warranty
# MANOVA, ANOVA and CVA for descriptive analysis data

library (agricolae)
library (candisc)

# we imported a file named ac10.csv which contained our descriptive
analysis data
da.d <- read.csv(file = "ac10.csv", header = TRUE, sep = ",", dec = ".")

head(da.d)
#   rep judge wine Confectionary GreenApple Lemonlime Grapefruit Peach
#1   1     1  w12           3.0        4.5       3.5        1.0   5.0
#2   1     2  w12           2.0        4.8       2.2        2.1   9.1
#3   1     3  w12           2.0        5.3       6.7        7.6   2.8
#4   1     4  w12           6.7        0.0       1.2        0.0   3.6
#5   1     5  w12           6.8        4.7       3.4        2.9   5.4
#6   1     6  w12           3.2        6.1       0.2        0.2   8.9

dim(da.d)
[1] 252  23
# this showed us that the data was correct since 12*3*7 = 252
# observations; 20 + 3 = 23 columns

# define judge, rep and wine as factors
# columns starting with a letter are automatically set as factors by R
# in our case judge and rep are not recognized as factors

is.factor(da.d$judge)       # is judge a factor?
#[1] FALSE

da.d$judge = as.factor(da.d$judge)
is.factor(da.d$judge)       # now it is
#[1] TRUE

is.factor(da.d$rep)         # is rep a factor?
#[1] FALSE

da.d$rep = as.factor(da.d$rep)
is.factor(da.d$rep)         # now it is
#[1] TRUE

is.factor(da.d$wine)         # wine is a factor as its entries start
[1] TRUE                     #  with a letter

# combine all attributes and define it as a matrix

da.a = as.matrix(da.d[,-c(1:3)])  # use all columns but the first
                                    three (wine, judge, rep)
head(da.a)
#     Confectionary GreenApple Lemonlime Grapefruit Peach Apricot
#[1,]           3.0        4.5       3.5        1.0   5.0     3.0
```

```
#[2,]           2.0         4.8        2.2         2.1    9.1      8.6
#[3,]           2.0         5.3        6.7         7.6    2.8      1.1
#[4,]           6.7         0.0        1.2         0.0    3.6      6.2
#[5,]           6.8         4.7        3.4         2.9    5.4      3.9
#[6,]           3.2         6.1        0.2         0.2    8.9      0.1


# create a 3-way MANOVA model with all 2-way interactions

da.lm = lm(da.a ~ (wine + rep + judge)^2 , data= da.d)

# run the MANOVA

da.maov = manova(da.lm)
summary(da.maov, test='Wilks') # print MANOVA table, see Table 3

# in this MANOVA wine, rep and judge and the interaction wine:judge and
# rep:judge are significant at p< 0.05.
# since wine was significant we continue with individual ANOVA's
# using the same lm model but now using the ANOVA output

da.aov = aov(da.lm)
summary(da.aov)  # print ANOVA tables for each attribute. See Table 4 as
                 # an example

# sign. wine effect at p< 0.05 for: Confectionary, Peach, Apricot,
# Tropfruit, Melon, Freshveg, Oak, Bitter.

# sign. wine interaction (p<0.05) with sign. wine effect: Greenapple
# (W:J), Honey(W:J, W:R), Butter (W::J), Vanilla (W:J), Straw (W:J), Char
# (W:J), Sweet (W:J)
# apply pseudomixed model for attributes with sign. wine interactions
# Fnew = (MS_wine)/(MS_wine interaction)
# determine critical F-value for wine effect and W:J interaction, df1= df
# wine and df2 = aovsum = summary(da.aov)

dfW = aovsum[[1]][1,1] # df for wine effect
dfWR = aovsum[[1]][5,1] #df for wine:replication interaction
dfWJ = aovsum[[1]][5,1] # df for wine:judge interaction
newF_critWR = qf(0.95, dfW, dfWR) # critical F value for pseudomixed
                                  # model of wine and W:R
newF_critWJ = qf(0.95, dfW, dfWJ) # critical F value for pseudomixed
                                  # model of wine and W:J

# Greenapple had a sign. W:J; calculate new F-value and test significance

newF_Greenapple = aovsum[[2]][1,3]/(aovsum[[2]][5,3])
newF_Greenapple > newF_critWJ # F remains significant at p<0.05
                              # thus Greenapple has a sign. wine effect
```

```
newF_Honey = aovsum[[10]][1,3]/(aovsum[[10]][5,4])
newF_Honey > newF_critWJ
newF_Honey = aovsum[[10]][1,3]/(aovsum[[10]][5,4])
newF_Honey > newF_critWR     # F remains significant at p<0.05
                             # thus Honey has a sign. wine effect

newF_Butter = aovsum[[11]][1,3]/(aovsum[[11]][5,3])
newF_Butter > newF_critWJ    # F remains significant at p<0.05 thus
                             # Butter has a sign. wine effect

newF_Vanilla = aovsum[[12]][1,3]/(aovsum[[12]][5,3])
newF_Vanilla > newF_critWJ   # F remains significant at p<0.05 thus
                             # Vanilla has a sign. wine effect

newF_Straw = aovsum[[13]][1,3]/(aovsum[[13]][5,3])
newF_Straw > newF_critWJ     # F is not significant at p<0.05 thus
                             # Straw has no sign. wine effect

newF_Char = aovsum[[15]][1,3]/(aovsum[[15]][5,3])
newF_Char > newF_critWJ      # F is not significant at p<0.05 thus
                             # Char has no sign. wine effect

newF_Sweet = aovsum[[18]][1,3]/(aovsum[[18]][5,3])
newF_Sweet > newF_critWJ     # F remains significant at p<0.05 thus Sweet
                             # has a sign. wine effect

# continue further analyses with only significant attributes (significant
# wine effect). See Table 5 for a summary.
# create data subset with significant attributes only from original data
# sign. attributes are confectionary, greenapple, peach, apricot,
# tropfruit, melon, freshveg, honey, butter, vanilla, oak, bitter and
# sweet


da.s = da.d[, c(3, 4, 5, 8:15, 17, 20:21)]
head(da.s)
is.factor(da.s$wine)   # Wine is a factor as its entries start
                         with a letter

# combine all attributes and define it as a matrix

da.as = as.matrix(da.s[,-1])
head(da.as)

# calculate LSDs or HSDs using the agricolae package
# install the agricolae package first from CRAN and then load it

library(agricolae)

# for LSD use LSD.test, individually for all your attributes of interest
```

```
# build lm models for each attr:

Confectionary.lm = lm(Confectionary ~ (wine + rep + judge)^2, data=da.d)
Confectionary.LSD = LSD.test(Confectionary.lm, trt='wine', group = TRUE)
print(Confectionary.LSD$statistics)
print(Confectionary.LSD)

GreenApple.lm = lm(GreenApple ~ (wine + rep + judge)^2, data=da.d)
GreenApple.LSD = LSD.test(GreenApple.lm, trt='wine', group = TRUE)
print(GreenApple.LSD$statistics)
print(GreenApple.LSD)

# continue for all the other significant attributes. See Table 6.

# for HSD use HSD.test, analog to LSD.test


# Now we want to run a CVA on the significant data set using the candisc
# package
# install candisc package first from CRAN and load it afterwards

library(candisc)

# build MANOVA model with significant attributes only, using only the wine
# effect. See Monrozier, R.; Danzart, M. (2001) Food Quality and
# Preference 12:393-406

da.s.lm = lm(da.as ~ wine, data = da.s)

da.s.lm = lm(da.as ~ wine, data = da.s) # run the one-way MANOVA

da.man1 = manova(da.s.lm)
summary(da.man1, test='Wilks') # print the one-way MANOVA table

da.cva = candisc(da.s.lm) # run the CVA
da.cva          # extract CVA output => eigenvalues,
                # variance ratios and Bartlett's test for sign.CVs
                # See Table 7.
plot(da.cva, type = 'n') # plots the CVA biplot, output not shown

# For a better looking plots with 95% confidence intervals use the
# following code
# We will plot the first two canonical variates. If you want to plot
#canonical variables 1 and 3 then change da.cva$means[,1:2] to
#da.cva$means[, c(1,3] Figures 1 through 4

plot(da.cva$means[,1:2],xlab='CV1, 69.2%', ylab='CV2, 11.9%', cex=1.1,
xlim=c(-2,2), ylim=c(-2,2),axes=FALSE, cex.lab=.9)
                #get the % variance ratio explained from the CVA outpu
axis(side = 1, at = c(-2,0,2), line=1, col='darkgray', cex.axis=.9)
```

```
axis(side = 2, at = c(-2,0,2), line=1, col='darkgray', cex.axis=.9)
abline(h=0,v=0, col='darkgray', lty=3, lwd=.8)
symbols(x=da.cva$means[,1], y=da.cva$means[,2],
circles=2/sqrt(table(da.cva$scores[,1])), add=TRUE, inches=FALSE, lty=2,
lwd=.5, fg='black')
text(da.cva$means[,1:2], row.names(da.cva$means), cex=0.6, pos=4,
col="blue")
plot(da.cva$structure[,1:2], pch='', xlab="CV1, 69.2", ylab="CV2, 11.9%",
xlim=c(-.6,.6), ylim=c(-.6,.6), axes=FALSE, cex.lab=.9)
                    #get the % variance ratio explained from the CVA output
Axis(side=1, at=c(-.6,0,.6), line=1, cex.axis=.9, col='gray')
Axis(side=2, at=c(-.6,0,.6), line=1, cex.axis=.9, col='gray')
arrows(0,0,da.cva$structure[,1],da.cva$structure[,2], length=0.1)
text(da.cva$structure[,1:2], labels=row.names(da.cva$structure),pos=4,
cex=.7)
```

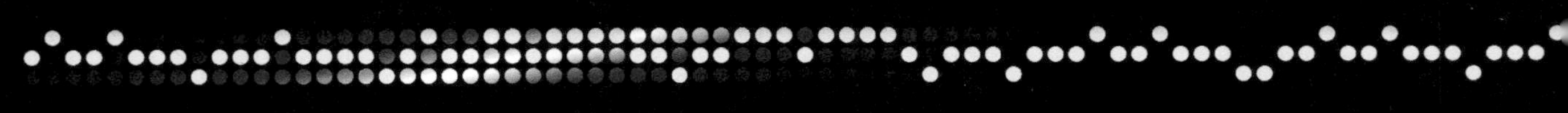

In this chapter

Objective

To determine how American mead (honey wines) differ from each other in terms of perceived color and other sensory attributes.

Case Study 3

A descriptive analysis[a] of American mead (honey wines)[b]

We used 14 mead wines for the study; due to limited amounts of each sample, we used a subset of 6 mead wines during the descriptive analysis (DA) training (Table CS3.1) (This is not the ideal situation since it is possible that one or more of the products used in the actual study may have attributes that were not found in the training subset. However, there are times when this is the only way to proceed.). During training, eight panelists (four women) determined, through consensus, which attributes described sensory differences among the samples. We then created reference standards to anchor these different terms. Table CS3.2 shows the reference standards for the significant attributes. The reference standards serve as a memory jogging device for the panelists (as in, that is what is meant by "astringent") as well as a "translation" device for anyone reading the research.

Once the panel was trained, they evaluated the 14 mead samples in triplicate in a completely randomized design. Within each session the panelist received seven samples. Samples (25 mL) were served in standard black ISO wine glasses covered with plastic Petri dish lids. All samples were coded with 3-digit random numbers. Panelists cleared their palates with water between sample evaluations and were asked to expectorate all samples. All evaluations, except color, were performed in individual climate-controlled sensory booths at 20°C and under white lights. The panelists used unstructured line scales to indicate the perceived intensities of each attribute. All data, except color, were collected using a computerized data system (FIZZ, Biosystèmes, Couternon, France).

[a]*Since this is the second case study involving descriptive analysis (DA), we will only describe the MANOVA and ANOVA results tables and instead of canonical variate analysis (CVA) we will use principal component analysis (PCA) as the multivariate technique.*

[b]*Based on unpublished data gathered by Ofelia Angulo, Ofelia Araceli Lopez Mejia, and Hildegarde Heymann.*

Sensory and Instrumental Evaluation of Alcoholic Beverages. http://dx.doi.org/10.1016/B978-0-12-802727-1.00009-0

Table CS3.1 Mead samples used for sensory evaluation

Identification code	Type[a]	Honey variety
RS*	Traditional honey wine	Orange blossom and wildflower
HM*	Sweet honey wine	N/A
MS	Dry mead	Clover
PR*	Mead	White clover
CS*	Mead	N/A
MT	Mead	N/A
BT	Mead	N/A
SM	Dry mead	Clover, alfalfa, and fireweed
LR	Reserve mead	N/A
SN*	Mead	N/A
RB*	Sweet mead	Jasmine and wildflower
SE	Traditional honey wine	Clover
VG	Mead	N/A
WL	Mead	N/A

[a]*Information on mead type and honey varieties were gleaned from the label but were not always available.*
Mead wines with an asterisk were used during training. All 14 mead wines were used during the evaluation phase.

3.1 Data analysis

The DA data were analyzed by three-way multi- and univariate analyses of variance (MANOVA and ANOVA, respectively) with main effects (mead M, panelist P, and replication R), as well as all two-way interaction effects ($M \times P$, $M \times R$, $P \times R$). In situations where the Mead by Panelist and/or Mead by Replication interaction effect(s) and the Mead main effect were significant a pseudomixed model was used to determine the importance of the interaction effect (see Case Study 2). Fisher's protected least significant differences (lsds) were used as the mean separation technique for the univariate ANOVAs. In all cases an alpha-value of 5% was used.

Principal component analyses (PCA) are usually performed on mean data; in this case if we had chosen to do a standard PCA we would have averaged the data across panelists and replications and would have used a 14 mead by 24 variable matrix. In many instances this would have been appropriate but in most cases we prefer to be able to see the effects of the panelist and replication variations in the multivariate data space, hence our usual choice of using CVA to describe the multivariate sensory descriptive data space (see Case Study 2).

Table CS3.2 Descriptors and reference standards used for the significant attributes in the sensory descriptive analysis (DA) of mead wine

Descriptor	References
Fruity aroma	Tutti-frutti flavor, Bubbaloo gum (Candylandstore.com)
Sulfury aroma	Hard-boiled egg—yolk only
Bruised apple aroma	Apple Cider, R.W. Knudsen's Family Farms, Chico, CA
Aged cheese	Grated Parmesan Cheese, Kraft Foods, Glendale, IL and Polder Gold Aged Goat Gouda, Holland
Musty aroma	Raw Button Mushroom, Nugget Market, Davis, CA
Nutty aroma	Mixed nuts (Blue Diamond, Sacramento, CA)
Almond aroma	1.5% Imitation Almond extract-water solution, McCormick, Hunt Valley, MD
Bitter taste	0.2 % Caffeine—water solution, Sigma-Aldrich, St. Louis, MO
Sweet taste	5% Sucrose—water solution, C&H Cane Sugar, San Francisco, CA
Sour taste	0.25% Citric Acid—water solution, Sigma-Aldrich, St. Louis, MO
Mulling spice flavor	Mulling spice mix, Williams Sonoma, San Francisco, CA
Bruised apple flavor	Apple Cider, R.W. Knudsen's Family Farms, Chico, CA
Fruity flavor	Tutti-frutti flavor, Bubbaloo gum (Candylandstore.com)
Almond flavor	1.5% Imitation Almond extract-water solution, McCormick, Hunt Valley, MD
Moldy flavor	Rind of Brie Cheese, Nugget Market, Davis, CA
Astringent mouthfeel	2% Alum—water solution, Sigma-Aldrich, St. Louis, MO

However, in SensomineR, an R-package created by Lê and Husson (2008) it is possible to perform a PCA on the *raw* sensory data (panellipse) which will then show 95% confidence intervals in the PCA space. The results of this specific version of a PCA are often very similar to that of a CVA. For practitioners who are more comfortable with PCAs, this is a good alternative to doing CVAs. We used this technique to determine the positions of the mead sample in the multidimensional space. Since our DA attributes were all measured on line scales of identical length we chose to use a covariance matrix as our input to the program (see Chapter 3).

3.2 Results

The panel evaluated 14 aroma terms and 6 (fruity, sulfury, bruised apple, aged cheese, musty, and almond aromas) were significantly different among the mead wines. All three taste attributes (sweet, sour, and bitter) differed significantly across the mead wines as did the mouthfeel attribute (astringency). The panel evaluated six retronasal flavor attributes and five (mulling spice, bruised apple, fruity, almond, and moldy flavors) differed significantly across the mead wines. Table CS3.3 shows the means and lsd values for each significant attribute. The sweet meads were CS, HM, LR, and Se while the driest ones were BT, MS, and SM. SM was very sour. The most bitter and astringent products were BT, MS, MT, and RB. The mead wines with the most fruity aroma were RS and PR while those with the most fruity flavor were CS and HM. For the bruised apple aroma, RS and LR were the most intense while HM had the most intense bruised apple flavor. RB had the most intense almond aroma and flavor, as well as the highest intensity of mulling spice flavor. The nutty aroma significantly discriminated among the samples with RB having the highest intensity, but the range (1.2–0.2) was extremely small. Meads with aromas and flavors that are typically considered less

Table CS3.3 Means and Fisher's protected least significant values for attributes differing significantly across mead wines

	Aroma							Taste
Mead	**Fruity**	**Sulfury**	**Bruised apple**	**Aged cheese**	**Musty**	**Nutty**	**Almond**	**Sweet**
BT	1.4 e	3.0 ab	2.0 e	1.2 b	2.9 ab	0.5 bc	0.5 cde	1.2 f
CS	2.2 cde	1.9 de	3.0 cd	1.1 b	2.8 ab	0.6 abc	0.4 de	7.7 a
HM	2.6 bc	0.6 g	2.5 de	0.3 c	1.0 e	0.2 c	1.0 bcd	6.2 b
LR	2.3 cde	0.9 fg	3.8 ab	0.9 bc	1.9 cd	1.0 ab	0.7 bcde	5.8 b
MS	2.0 cde	2.2 bcd	2.1 e	0.9 bc	2.5 bc	0.5 bc	0.5 de	1.1 f
MT	1.9 cde	1.8 def	2.1 e	0.8 bc	2.3 bc	0.3 c	0.8 bcde	2.0 e
PR	3.5 ab	1.8 def	3.0 cd	0.8 bc	2.5 bc	0.4 c	0.9 bcde	4.1 c
RB	2.5 cd	0.7 g	3.0 cd	0.2 c	1.5 de	1.2 a	2.1 a	3.5 cd
RS	4.3 a	0.5 g	4.4 a	0.6 bc	1.2 de	0.3 c	1.3 b	3.9 c
SE	2.1 cde	1.1 efg	3.5 bc	0.6 bc	1.9 cd	0.7 abc	1.2 bc	6.5 b
SM	1.5 de	2.9 abc	2.0 e	2.9 a	2.8 ab	0.3 c	0.6 bcde	1.7 ef
SN	1.5 e	1.9 def	1.9 e	0.9 bc	2.3 bc	0.5 bc	0.5 cde	3.6 cd
VG	1.4 e	3.4 a	1.9 e	1.3 b	3.5 a	0.2 c	0.7 bcde	4.1 c
WL	1.9 cde	2.1 cde	2.2 de	0.9 bc	2.8 ab	0.5 bc	0.3 e	2.9 d
lsd	0.98	0.94	0.78	0.82	0.87	0.60	0.68	0.79

Table CS3.3 Means and Fisher's protected least significant values for attributes differing significantly across mead wines (*cont.*)

	Taste							Mouthfeel
Mead	**Sour**	**Bitter**	**Bruised apple**	**Mulling spice**	**Fruity**	**Almond**	**Moldy**	**Astringent**
BT	3.7 bc	6.5 a	1.5 d	2.2 b	0.4 f	1.3 cde	2.0 ab	4.3 ab
CS	2.6 de	1.1 h	4.8 a	0.6 e	4.1 a	0.8 def	0.3 g	1.0 f
HM	3.5 bcd	2.4 fg	4.2 ab	3.0 a	3.6 ab	1.0 de	0.4 g	2.7 cd
LR	2.8 bcde	4.4 cd	4.0 b	1.5 cd	1.6 cd	2.3 ab	0.3 g	3.7 b
MS	3.0 bcde	5.5 ab	2.3 c	2.1 bc	0.8 ef	1.0 de	1.7 bc	3.8 ab
MT	2.9 bcde	5.7 a	2.5 c	2.0 bc	1.1 def	1.3 cde	1.5 bcd	3.8 ab
PR	3.8 b	2.4 fg	3.7 b	2.0 bc	2.3 c	0.7 ef	1.3 cde	2.5 cde
RB	2.8 cde	5.6 ab	2.8 c	2.4 ab	1.2 def	2.9 a	0.8 efg	4.6 a
RS	2.7 de	4.6 bc	4.5 ab	1.9 bc	1.5 cde	1.9 bc	1.0 def	2.8 c
SE	2.3 e	1.5 gh	4.8 a	2.3 ab	3.2 b	0.9 de	0.4 fg	1.7 ef
SM	6.3 a	3.0 ef	2.2 cd	1.0 de	1.1 def	0.2 f	2.6 a	2.7 cd
SN	2.4 e	3.9 cde	2.7 c	1.5 cd	1.4 de	1.4 cde	1.1 de	2.8 cd
VG	2.4 e	2.1 fg	3.7 b	1.8 bc	1.8 cd	1.3 cde	1.2 cde	2.0 de
WL	3.0 bcde	3.6 de	2.8 c	2.3 ab	1.4 de	1.4 cd	1.2 cde	2.5 cde
lsd	0.96	1.00	0.79	0.68	0.80	0.70	0.63	0.82

Products with the same letter within a column do not differ significantly.

pleasant were VG and BT (sulfury aroma); VG (musty aroma); SM (aged cheese aroma); and SM and BT (moldy flavor).

Fig. CS3.1 provides the loadings for the first two principal components (PCs) which explained 82% of the variance in the data space. The third PC explained an additional 6% (data not shown) and did not affect the interpretation of the first two dimensions. The loadings plot shows the attribute vectors and attributes that are highly positively correlated have small positive angles between their vectors. Highly negatively correlated attributes have small negative angles between vectors and vectors that are at 90 degree angles are not correlated (orthogonal). Additionally, attribute vectors with small positive angles relative to a PC are highly correlated to that axis. Short attribute vectors are not important in explaining the multidimensional space while long ones are very important.

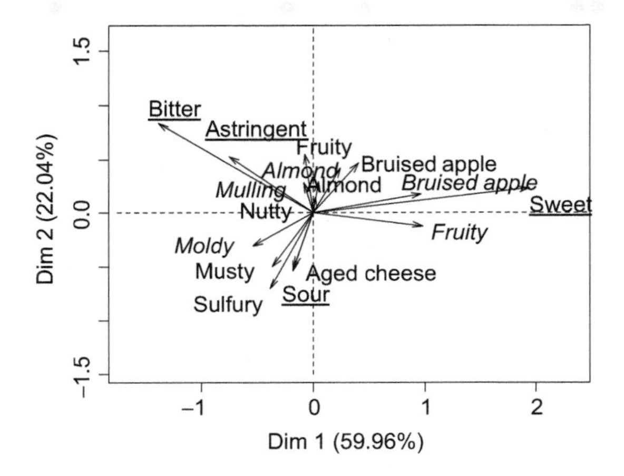

Figure CS3.1
Principal component analysis (PCA) loadings plot of PC 1 and PC 2 for the 14 mead wines
Only attributes that significantly discriminated among the samples ($p < 0.05$) were used in the calculation. PCA was performed on the covariance matrix. Aroma attributes are *normal font* and flavor by mouth (retronasal) attributes are *italicized.* Taste and mouthfeel attributes are *underlined.*

Sweet taste, bruised apple and fruity flavors are highly loaded in the positive direction of PC1. Bitter taste and astringent mouthfeel are at approximately 45 degree to both PC1 and PC2 and are thus important in describing both axes. Bitter and astringent are highly correlated and it is possible that the panel used both terms very similarly. Aged cheese aroma and sour taste are highly correlated and loaded in the negative direction of PC2. Fruity aroma and, to a lesser extent, bruised apple aroma loaded in the positive direction of PC2. Attributes describing less pleasant aromas and flavor (musty and sulfury aromas and moldy flavor) are somewhat correlated with one another and are also somewhat loaded on both PC1 and PC2. Almond aroma and flavor, nutty and mulling spice flavors have very short vectors and do not add much information about the space.

In Fig. CS3.2, mead wines (CS, SE, and HM) on the right side of PC1 are higher in perceived sweetness, fruity and bruised apple flavors than those on the left side (BT, MS, and MT) and a quick check of Table CS3.3 will show general agreement. SM is different from all other meads in the space and is quite sour, with an intense aged cheese aroma and some musty aroma and a moldy flavor. RB is the most bitter and astringent mead and is significantly different than all the other samples in the space. PR and VG are not significantly different from each other; WL and SN are also quite similar and SN shares some similarities with LR.

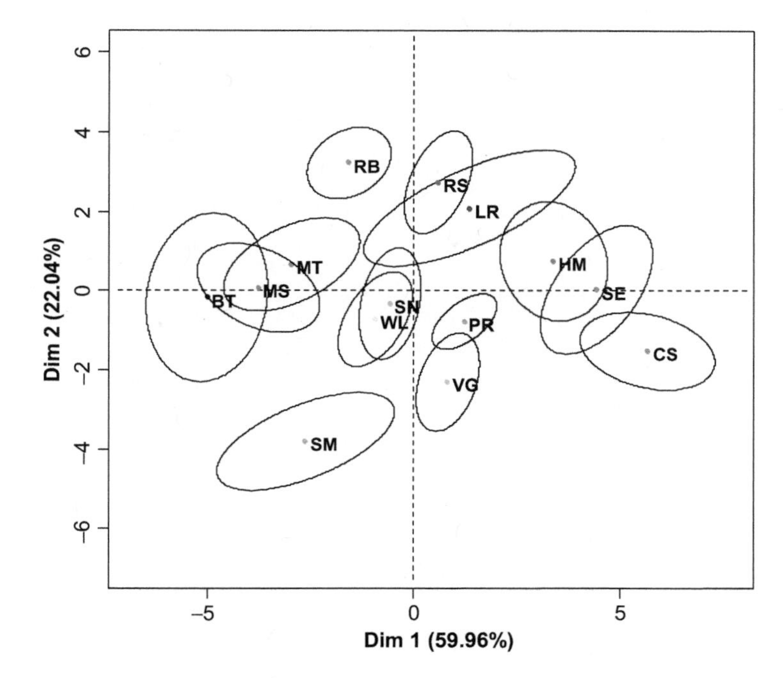

Figure CS3.2 PCA scores plot of PC 1 and PC 2 for the 14 mead wines
Ninety-five percent confidence ellipses are indicated. See Table CS3.1 for codes.

3.3 Conclusions

The mead wines tested in this study differed in their sensory attributes. In general, some meads were very sweet (CS, SE, and HM) while others were dry (BT, MS, and MT). One of the mead wines had high intensities of less pleasant aromas and flavor and was very sour (SM).

Reference

Lê, S., Husson, F., 2008. SensoMineR: a package for sensory data analysis. J. Sens. Stud. 23, 14–25.

```
# all code comes without any warranty
# We imported a file named Mead.csv
# 8 judges, 3 replicates, 14 wines, 24 sensory attributes
# We will only show some of the code for the MANOVA and ANOVAs, since it
# is very similar to the code shown in Case Study 2.
library (SensoMineR)

# we imported a file named Mead.csv which contained our descriptive
analysis data
da.d <- read.csv(file = "Mead.csv", header = TRUE, sep = ",", dec = ".")

da.d=Mead
head(da.d)
dim(da.d)              # 8*3*14 = 336 observations; 24 + 3 = 27 columns

# define judge, rep and mead as factor
# columns starting with a letter are automatically set as factors by R
# in our case judge and rep are not recognized as factors

is.factor(da.d$CJ)                                    # Is judge a factor?
da.d$judge = as.factor(da.d$CJ)
is.factor(da.d$judge)                                 # Now it is

is.factor(da.d$NR)                                    # Is rep a factor?
da.d$rep = as.factor(da.d$NR)
is.factor(da.d$rep)                                   # Now it is

is.factor(da.d$ProductName)
da.d$mead = as.factor(da.d$mead)
      is.factor(da.d$mead)   # Mead is a factor as it sentries start with
                                                      a letter

# combine all attributes and define it as a matrix
da.a = as.matrix(da.d[,-c(1:3)])           # use all columns but the first
                                           # three (judge, rep, mead)
head(da.a)

# building a 3-way MANOVA model with all 2-way interactions
da.lm = lm(da.a ~ (mead + rep + judge)^2 , data= da.d)

# run the MANOVA

da.maov = manova(da.lm)
summary(da.maov, test ="Wilks") # print MANOVA table
da.aov = aov(da.lm)
summary(da.aov)  # print ANOVA tables for each attribute
                 # the rest of the code for the means separation technique
                 # would be similar to the code in Case Study 2 and is not
                 # shown here

# Now we use the raw sensory DA data to perform a PCA

# we create a new data frame with the judge, replication and mean columns
# as well as the significant attributes
da.as = (da.d[,-c(4, 7:9, 13:16, 26)])
head(da.as)

panellipse(da.as, col.p = 3, col.j =1, firstvar =4, alpha = 0.05,
           coord = c(1,2), scale.unit = FALSE, nbsimul = 500,
           nbchoix = NULL, group = NULL, name.group = NULL,
           level.search.desc = 0.2, centerbypanelist = TRUE,
```

```
            scalebypanelist = FALSE, name.panelist = FALSE,
            variability.variable = TRUE, cex = 1, color = NULL)

# we perform the analysis on the significant attribute data frame. Our
# panelist information is in column 1 (col.j=1) and our mead information
# is in column 3 (col.p=3). Our first attribute is in column 4
#(firstvar=4) and we are calculating the first two principal components
# (coord = c(1,2). We want to use a covariance matrix (scale.unit = FALSE)
# and we are running 500 simulations to determine the 95% confidence
# ellipses. See figures 1 and 2
```

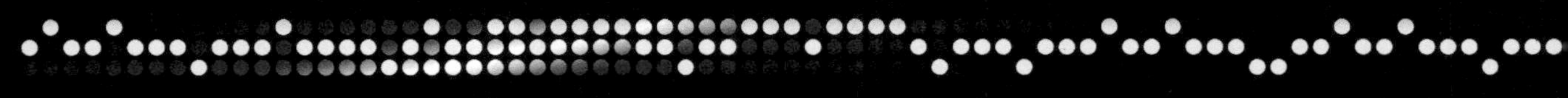

In this chapter

Objective

We were interested in characterizing the sensory attributes of Californian blanc de blanc sparkling wine 1 min and 5 min after pouring the wine into a sparkling wine flute.

Case Study 4

A descriptive analysis[a] of Californian sparkling wines[b]

4.1 Sample considerations

Sparkling wine, beer, sparkling cider, and a number of other alcoholic (and nonalcoholic) beverages contain dissolved carbon dioxide (CO_2) and the bubbles/foam created when the pressure in the container is released (Vreme et al., 2015) are an integral part of the sensory experience of consuming the beverage. However, making sure that each panelist receives a beverage with the same (or at least similar amounts) of carbonation is not easy. As said by Kemp et al. (2015) "studies lack attention to glass type, sample order effect on foam and flavor, and the time between opening bottles and then pouring and imbibing." Many researchers simply do not address the carbonation issue in their studies. For example Rey-Salgueiro et al. (2013) working with sparkling mineral waters, Kappes et al. (2006, 2007) working with cola and lemon/lime beverages, and Pérez-Magariño et al. (2015) working with sparkling wines make no mention of the potential issue at all, and thus do not provide details how these carbonated samples were evaluated and how the study was designed to overcome the temporal changes due to the presence of CO_2. In an earlier study Martínez-Lapuente et al. (2013) did partially address the issue by pouring the wine into the glass slowly to avoid air bubble formation (whatever that means). However, they did not correct for serving order or time in the glass prior to imbibing.

Lorjaroenphon and Cadwallader (2015) working with cola beverages poured 50 mL into 125 mL Nalgene sniff bottles and kept the bottles open at room temperature for 30 min prior to evaluation to allow most of the carbon dioxide to passively escape. Similarly Martin (2002) actively degassed Champagne in

[a] *Since this is the third case study involving descriptive analysis (DA), we will only supply the R-code for the ANCOVA.*
[b] *Based on data used in Hood White and Heymann (2015).*

Sensory and Instrumental Evaluation of Alcoholic Beverages. http://dx.doi.org/10.1016/B978-0-12-802727-1.00010-7

a blender for 4 min. However, in carbonated beverages the carbon dioxide is an integral part of the sensory profile and removing the "bubbles" irretrievably changes this profile, additionally stirring or active blending will also lead to other volatile compounds being lost.

On the other hand, for their consumer study, Leksrisompong et al. (2013) served a lemon/lime carbonated drink at 6°C immediately after pouring about 90 mL into Styrofoam cups; 2 min after opening, any remaining sample in the opened cans or bottles was then discarded to prevent variability in the loss of carbonation from influencing the results. For the descriptive analysis of these beverages the same authors (Leksrisompong et al., 2012) poured 88 mL carbonated beverage into a 118 mL plastic Solo soufflé cup which was then lidded. They then waited 2 min before the panelists started their evaluation since this procedure "minimize(d) perceived carbonation variability." This is somewhat problematic since it is not clear that the consumer and descriptive panels evaluated the same samples.

De La Presa-Owens et al. (1998) waited 5 min after pouring sparkling wines to minimize carbonation variability. These are a few of the many examples existing in the literature. Thus one of our goals for this case study was to show an example where the carbonation issue was explicitly addressed and evaluated by comparing the sensory profiles of the sparkling wines 1 and 5 min after pouring.

4.2 Materials and methods

For this case study we are using five of the eight Californian traditional method blanc de blanc sparkling wines (SW), each containing at least 75% Chardonnay, used in the original study (Table CS4.1).

Panelists were trained over six 1-h training sessions. SW bottles were opened once all panelists arrived to the given session. During the training period each SW was considered by the panelists at least once. However, since we had only a few bottles of some SW we used an additional sparkling wine for training purposes. This wine was very similar to the assessed SW, as it contained at least 75% Chardonnay, was made by the traditional method and was a nonvintage Colombia Valley, Washington SW. Additionally, prior to training, an experienced tasting group evaluated this SW and deemed it to have similar attributes to the sparkling wines in the study. Panelists generated all attributes and reference standards via consensus and chose to use 16 aroma, 4 taste and 2 bubble mouthfeel attributes (Table CS4.2).

Panelists were introduced to timed assessments from the second training session onward and by the last training session they could assess the aroma attributes in 1 min, with the taste and bubble attributes taking another minute. A timer was provided on the screen with the attributes. Panelists were instructed not to swirl glasses during the aroma assessment. Each panelist adjusted their seat to a height

Table CS4.1 Sparkling wines, retail price, and vintage

Wine	Price (US$[a])	Vintage	Codes used in Fig. CS4.2 [b]
Piper Sonoma (W1)	**17.99**	**NV[c]**	**1.1, 1.5**
Gloria Ferrer (W2)	32.00	2007	
Mumm Napa (W3)	38.00	2007	
Schramsberg (W4)	**37.00**	**2009**	**4.1, 4.5**
Korbel (W5)	**10.99**	**NV**	**5.1, 5.5**
Frank Family (W6)	**45.00**	**2008**	**6.1, 6.5**
Iron Horse (W7)	40.00	2007	
Chandon Napa (W8)	**30.00**	**NV**	**8.1, 8.5**

[a] *Per 750 mL bottle.*
[b] *The number after the period indicates the number of minutes since the wine was poured.*
[c] *Nonvintage.*
Wines used in this case study are set in boldface. In Table CS4.1 and in all data output the original numbering for the complete set of 10 sparkling wines has been retained. This was done to make it easier for anyone interested in comparing the results of the subset to the entire set of wines.

Table CS4.2 Sensory attributes and reference standards used in the sensory descriptive analysis of Californian sparkling wines

Attribute	Reference standard[a]
Overall aroma	Defined as the total intensity of the perceived aroma
Apple aroma	12 g chopped green apple, 15 mL water, 1/4 t ascorbic acid (King Arthur Flour, Norwich, VT)
Citrus aroma	3 g orange peel, 2.5 g grapefruit peel, 1 g lemon peel, 15 mL wine
Tropical aroma	10 mL pineapple juice (Dole, Thousand Oaks, CA), 5 drops guava juice, 1 drop passion fruit juice (last two juices from Kern's Beverages, Santa Ana, CA)
Stone fruit aroma	10 mL peach juice (Kern's Beverages), 4 drops liquid from canned apricots (Del Monte Foods, San Francisco, CA)
Artificial fruit aroma	One Jolly Rancher Watermelon flavor (The Hershey Company, Hershey, PA), 1 drop banana extract (McCormick and Company, Baltimore, MA), 30 mL wine
Confectionary aroma	1/4 t honey (Honey Bear, Dutch Gold Honey, Lancaster, PA), 15 mL wine
Floral aroma	15 mL wine, 3 drops rose extract (Star Key White In, Congers, NY), 4 drops orange blossom extract (Eden Botanicals, Petaluma, CA), 4 drops violet essence (Uncle Roy's Comestible Concoctions, Moffat, Scotland), 5 drops linalool essence (Eden Botanicals)

(*Continued*)

Table CS4.2 Sensory attributes and reference standards used in the sensory descriptive analysis of Californian sparkling wines (*cont.*)

Attribute	Reference standard
Yeasty/bready aroma	*Standard 1*: two 1 cm cubes toasted brioche (Nugget Brioche Rolls, Nugget Markets, Woodland, CA) *Standard 2*: 0.25 g Fleischman's Baker's Yeast (ACH Foods, Oakbrook, IL), 15 mL water
Nutty/oxidized aroma	*Standard 1*: four roasted almonds (Nugget Mixed Roasted, No Peanuts, Nugget Markets, Woodland, CA), 30 drops Pedro Ximenez Sherry (Hidalgo, Spain) *Standard 2*: 15 mL sherry (Tio Pepe, Spain)
Cooked apple aroma	6 g dried cut red apples (Trader Joe's, Monrovia, CA), 1 T unsweetened apple sauce (Motts, Dr Pepper-Snapple, Plano, TX)
Woody/toasty aroma	0.5 g French oak small chips, HT (ēVoak, Oak Solutions Group, Napa, CA), additional toasting with a blow torch until evenly browned, then added to 15 mL water
Herbal aroma	0.15 g five herb seasonings blend containing basil, chervil, tarragon, marjoram, chives (Davis Co-Op, Davis, CA)
Canned veg aroma	25 mL artichoke juice (Reese Specialty Foods, Nashville, TN), 10 mL canned green bean juice (Jolly Green Giant, General Mills, Minneapolis, MN), 20 drops fresh onion juice
Chemical aroma	*Standard 1*: 30 drops ethyl acetate (Sigma Aldrich, St Louis, MO), 15 mL wine *Standard 2*: standard 1 + 1/2 t white vinegar (HJ Heinz Co., Pittsburg, PA), 30 mL water
Sulfur aroma	*Standard 1*: 20 drops SO_2, 15 mL wine *Standard 2*: four cut rubber bands (Amazon, Seattle, WA), 30 mL wine
Sour taste	1.5 g/L Citric acid (Sigma Aldrich, St Louis, MO)
Sweet taste	*Standard 1*: 10 g/L sucrose *Standard 2*:15 g/L sucrose *Standard 3*: 20 g/L sucrose (C&H Sugar, Crockett, CA)
Viscous mouthfeel	*High concentration*: 3 g/L carboxymethyl cellulose (CMC). *Low concentration*: 1.5 g/L CMC (CMC, Sigma Aldrich, St Louis, MO)
Bitter taste	800 mg/L caffeine (Sigma Aldrich, St Louis, MO)
Bubble size—mouthfeel	*Low*: gently sparkling mineral water (Hildon Estate, Test Valley, Hampshire, England)
Bubble concentration—mouthfeel	*Low/med*: mineral water (San Pellegrino, Lombardy, Italy) *Med/high*: mineral water (Perrier, Vergeze, France) *High*: spring water (Mountain Valley Spring Water, Hot Springs, AR)

[a]All "wine" listed refers to Franzia Vintner's Select Chardonnay (Ripon, CA).

where their nose fit directly over the glass, requiring no movement or agitation of the glass.

Formal evaluation sessions were held for over 6 days and the panelists attend one of two sessions (10 a.m. or 2 p.m.) offered on each day. All panelists were required to smell aroma reference standards prior to each session. Formal evaluations took place at ambient temperature (21°C) in individual booths and under red light to mask visual differences. At the start of each session, panelists were asked to adjust their seats to accommodate the aforementioned aroma assessment technique. Panelists were required to expectorate all samples and were given unsalted saltine crackers (Nabisco, East Hanover, New Jersey) and water as palate cleaners in between samples.

All wines, at both 1- and 5-min time points, were presented in a forced randomized order, in triplicate; such that all paired combinations of wines were assessed. Each wine was opened immediately prior to pouring, and only when panelists were already seated in the evaluation booths. Unstructured line scales anchored by none and high were used and data were recorded using FIZZ (Version 2.47B, Biosystèmes, Couternon France). There were no forced answers; panelists only rated attributes that they perceived. To accomplish this, FIZZ software was programed to default to a zero rating for all attributes. Panelists were instructed to only adjust the cursor placement on the line scale for each attribute they would rate above zero, thereby, eliminating the need for each panelist to adjust the cursor for each attribute, unless needed. This was done to allow the panelists to complete the evaluation within the 1 min time frame for aroma, and taste and bubble characteristics, respectively.

Serving: Panelists were asked to assess eight wines in each formal session. All wines were presented at room temperature (21°C). No panelist assessed the same wine at two different time points in the same session, with the goal of randomizing order effects. Wines to be assessed at 5 min were poured immediately after the bottle was opened, and timed. Wines to be assessed at 1 min were then poured and given to panelists. Wines were presented to the panelists monadically.

Pouring and weights: Due to the interference of CO_2, wines volumes could not be controlled while maintaining the integrity of the experimental design. To combat this, the weight of each sample was recorded and used in the data analysis. Each empty glass was tared, and then weighed with wine. All wines were poured with the glass held at a 45 degree angle, with the wine poured down the side. Each weight was recorded and associated to the specific wine, at the given time point, for each panelist and replicate because we did not know whether the different weights would affect any of our sensory attributes. Recording the weight information allowed us to use weight as a covariate in our later data analyses. For wines evaluated at 1 min, the time lapse between pouring and assessment was less than 10 s.

4.3 Data analyses

Two panelists each missed one formal session of the descriptive analysis resulting in imputation of 2.7% of the data. Imputations were conducted by averaging scores of missed samples with the data from the two replications that each panelist was present. To differentiate the two time points for one wine, each time point was treated as an independent product. Thus, Wine 1 at 1 min was analyzed as a different product than Wine 1 at 5 min.

Recorded wine weights were treated as covariates in the univariate three-way analyses of covariance (ANCOVA, main effects: judge, wine and replication and all two-way interactions) of all sensory attributes. A significance level of alpha = 0.05 was used for all analyses. A pseudomixed model was calculated for significant attributes with both a significant wine effect and significant wine interactions (judge by wine and wine by replicate; Naes and Langsrud, 1998). Fisher's least significant difference (lsd) was calculated for all significant wine attributes.

The R-package SensoMineR and its functions principal component analysis (PCA) and panellipse were used to graphically display sample differences based on significant attributes using *raw* data. Additionally, we also used the R-package candisc to do a canonical variate analysis (CVA) of the significant attributes, again on *raw* data. The 95% confidence circles were calculated using the method of Chatfield and Collins (1980). We did both panellipse and CVA to show that the two methods give essentially the same results.

4.4 Results

Based on the ANCOVA none of the sensory attributes were affected by the pouring weights. This is in contrast to the publication (Hood White and Heymann, 2015) where with all the wines we found that only bubble concentration was significantly affected by pouring weight. In that case the pours with larger weights were found to have a greater perceived mouthfeel bubble concentration. All other attributes were unaffected by pour weight. The average pour weight for the study was 22.9 ±3.7 g. We then proceeded with our normal (as described in Case Study 2) MANOVA followed by univariate ANOVAs and based on this seven aroma, one taste and two bubble attributes were significantly different across wines and times (Table CS4.3). In general, at 1 min all sparkling wines were perceived to be high in bubble concentration, bubble size, and sour taste, while at 5 min all sparkling wines were significantly lower in perceived bubble size and bubble concentration. There is a trend for wines to increase in perceived confectionary and nutty aromas and to a lesser extent in yeasty bread aromas from 1 to 5 min. Bitterness, apple, floral and cooked apple aromas did not change over time and the same is true for citrus aroma with the exception of sparkling wine 4 that decreased significantly in citrus aroma from 1 to 5 min.

In Case Study 2 we used CVA to create a two-dimensional data space for the DA data and in Case Study 3 we used the panellipse function in the R-package

Table CS4.3 Mean values and lsds for significant sensory attributes

Wine	Aroma attributes						
	Apple	Citrus	Conf	Floral	Yeast bread	Nutty oxidized	Cooked apple
1.1	2.0 a	1.3 a	1.2 de	1.9 a	1.2 c	1.2 bcd	1.0 ab
1.5	1.6 a	1.1 ab	2.5 abc	1.4 ab	1.2 bc	1.0 cd	0.5 b
4.1	1.5 a	1.4 a	1.7 cde	0.6 d	1.1 c	1.6 bcd	1.1 ab
4.5	1.8 a	0.5 bc	2.4 abc	0.6 cd	1.4 bc	2.1 ab	1.70 a
5.1	1.9 a	0.6 bc	1.4 de	1.2 abcd	0.9 c	0.9 d	1.7 a
5.5	1.7 a	0.7 abc	1.7 cde	1.3 abc	1.1 c	1.1 bcd	1.6 a
6.1	1.3 ab	0.4 c	.9 e	0.9 bcd	2.0 ab	1.8 abcd	1.2 ab
6.5	0.6 b	0.4 c	2.6 ab	0.8 bcd	2.5 a	2.7 a	1.2 ab
8.1	1.2 ab	0.7 abc	2.1 bcd	1.1 bcd	1.5 bc	2.0 abc	1.1 ab
8.5	1.3 ab	0.3 c	3.1 a	0.7 bcd	1.5 bc	2.7 a	1.7 a
lsd	0.91	0.68	0.94	0.70	0.78	1.07	0.90

Wine	Taste attribute	Mouthfeel attributes	
	Bitter	Bubble size	Bubble concentration
1.1	2.2 ab	3.3 bc	4.7 a
1.5	1.9 b	2.4 d	4.4 ab
4.1	2.4 ab	4.3 a	5.0 a
4.5	2.4 ab	2.9 bcd	3.8 bc
5.1	2.6 a	3.6 ab	5.0 a
5.5	2.2 ab	2.4 d	3.3 cd
6.1	2.4 ab	3.8 ab	5.0 a
6.5	2.4 ab	2.5 cd	2.8 d
8.1	2.6 a	3.6 ab	4.7 a
8.5	2.8 a	2.4 d	2.8 d
lsd	0.71	0.88	0.82

The first number is the sparkling wine code and the second is the time of evaluation with 1 = 1 min after pouring and 5 = 5 min after pouring. Conf, Confectionary; lsd, least significant difference.
Means with the same letter do not differ significantly at $p<0.05$.

SensoMineR to do the same. In this case study we used both methods (CVA and the panellipse function in SensoMineR) to show the reader that the results from the two methods are usually quite similar and that either method could be used.

Fig. CS4.1 shows the CVA scores and loadings plots for the data space of the significant attributes and Fig. CS4.2 shows the PCA scores and loadings plots for the same data.

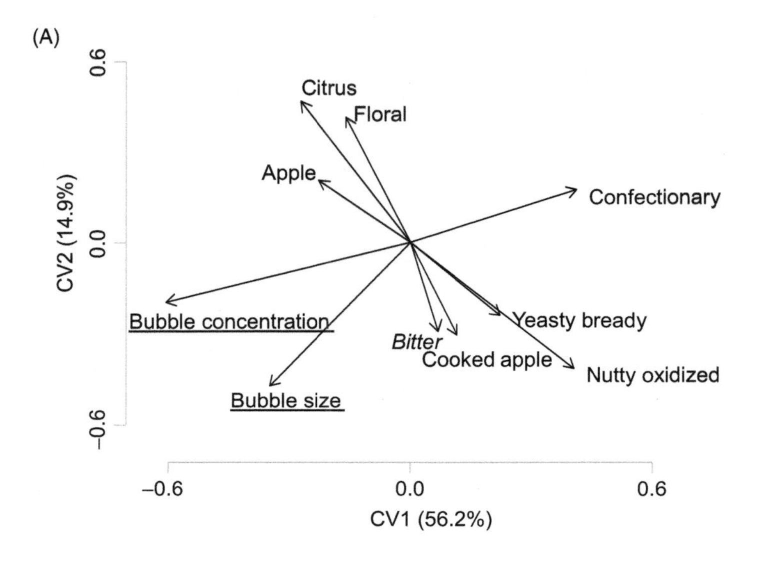

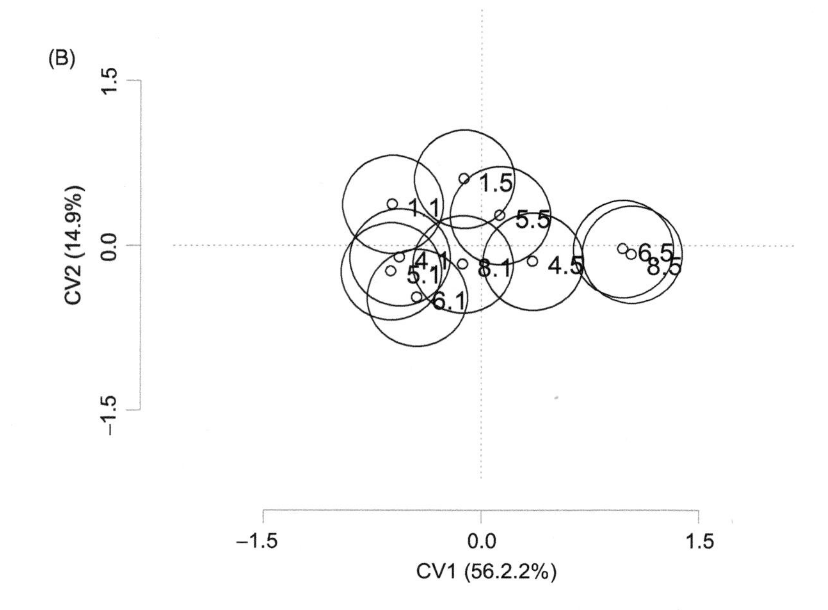

Figure CS4.1 (A) Canonical variate analysis (CVA) loadings plot of CV 1 and CV 2 for the sparkling wines. Only attributes that significantly discriminated among the samples ($p < 0.05$) were used in the calculation. Aroma attributes are *normal font*, taste attributes are *italicized*, and visual attributes are *underlined*. (B) CVA scores plot of CV 1 and CV 2 for the sparkling wines. Ninety-five percent confidence circles are indicated. See Table CS4.1 for codes.

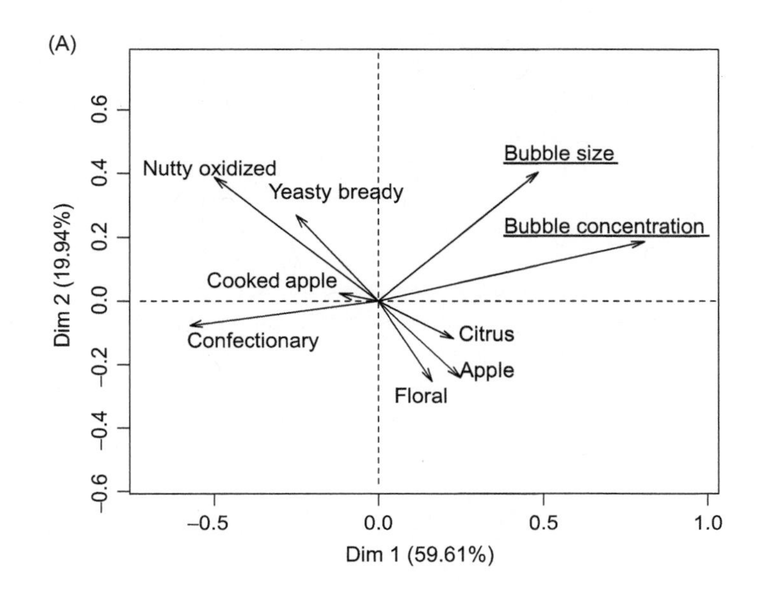

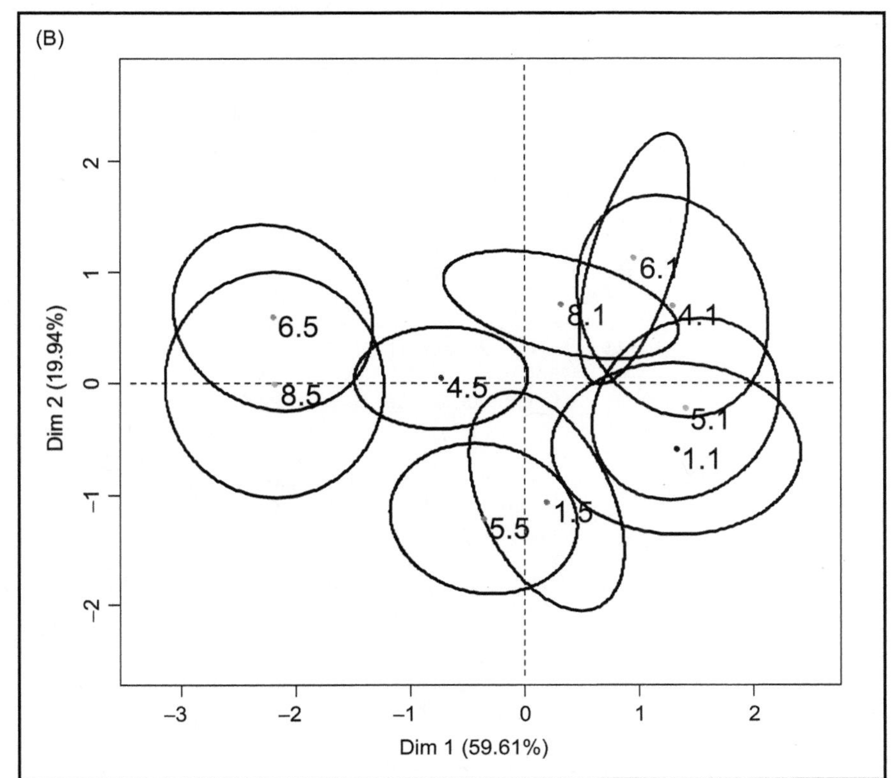

Figure CS4.2
(A) Principal component analysis (PCA) loadings plot of PC 1 and PC 2 for the sparkling wines. Only attributes that significantly discriminated among the samples ($p < 0.05$) were used in the calculation. PCA was performed on the covariance matrix. Aroma attributes are *normal font,* and visual attributes are *underlined.*
(B) PCA scores plot of PC 1 and PC 2 for the sparkling wines. Ninety-five percent confidence ellipses are indicated. See Table CS4.1 for codes.

The first two dimensions of the CVA space explained 71.1% of the variance ratio in the data set while the first two dimensions of the PCA space explained 79.55% of the variance in the data set. The confidence circles of the CVA were all same size since we used the method of Chatfield and Collins (1980) whereas the ellipses of the PCA space varied based on the underlying variation in the data for each sample. Thus for wine 4.5 the variation in the first two dimensions was much smaller than say for wine 1.1. Based on this information the PCA representation is likely to be more accurate. However, the major conclusions would be the same regardless of using the CVA or the PCA. Specifically, all wines, except wine 1 were significantly different at time point 1 min and time point 5 min. The wines, again all except wine 1, at the 1 min time point had more visual bubbles and higher bubble concentrations. Wine 6 and 8 at 5 min had higher confectionary and nutty oxidized aromas.

4.5 Conclusions

In this case study we showed the reader how to do an analysis of covariance and we also compared the results of creating the multivariate space with a CVA or using the panellipse function in SensoMineR. We also showed that the time elapsed from opening a sparkling wine until assessment significantly changes the sensory properties of the wines.

References

Chatfield, C., Collins, A.J., 1980. Introduction to Multivariate Analysis, first ed. Chapman and Hall, London, pp. 189–210.

De La Presa-Owens, C., Schlich, P., Davies, H.D., Noble, A.C., 1998. Effect of Méthode Champenoise Process on aroma of four *V. vinifera* varieties. J. Enol. Vitic. 49, 289–294.

Hood White, M., Heymann, H., 2015. Assessing the sensory profiles of sparkling wine over time. Am. J. Enol. Vitic. 66, 156–163.

Kappes, S.M., Schmidt, S.J., Lee, S.-Y., 2006. Descriptive analysis of cola and lemon/lime carbonated beverages. J. Food Sci. 71, S583–S589.

Kappes, S.M., Schmidt, S.J., Lee, S.-Y., 2007. Relationship between physical properties and sensory attributes of carbonated beverages. J. Food Sci. 72, S1–S11.

Kemp, B., Alexandre, H., Robillard, B., Marchal, R., 2015. Effect of production phase on bottle-fermented sparkling wine quality. J. Agric. Food Chem. 63, 19–38.

Leksrisompong, P.P., Lopetcharat, K., Guthrie, B., Drake, M.A., 2012. Descriptive analysis of carbonated regular and diet Lemon-Lime beverages. J. Sens. Stud. 27, 247–263.

Leksrisompong, P.P., Lopetcharat, K., Guthrie, B., Drake, M.A., 2013. Preference mapping of lemon lime carbonated beverages with regular and diet beverage consumers. J. Food Sci. 78, S320–S328.

Lorjaroenphon, Y., Cadwallader, K.R., 2015. Characterization of typical potent odorants in cola-flavored carbonated beverages by aroma extract dilution analysis. J. Agric. Food Chem. 63, 769–775.

Martin, N., 2002. Sweet/sour balance in champagne wine and dependence on taste/odour interactions. Food Qual. Prefer. 13, 295–305.

Martínez-Lapuente, L., Guadalupe, Z., Ayestarán, B., Ortega-Heras, M., Pérez-Magariño, S., 2013. Sparkling wines produced from alternative varieties: sensory attributes and evolution of phenolics during winemaking and aging. Am. J. Enol. Vitic. 64, 39–49.
Naes, T., Langsrud, Ø., 1998. Fixed or random assessors in sensory profiling? Food Qual. Prefer. 9, 145–152.
Pérez-Magariño, S., Martínez-Lapuente, L., Bueno-Herrera, M., Ortega-Heras, M., Guadalupe, Z., Ayestarán, B., 2015. Use of commercial dry yeast products rich in mannoproteins for white and rosé sparkling wine elaboration. J. Agric. Food Chem. 63, 5670–5681.
Rey-Salgueiro, L., Gosálbez-García, A., Pérez-Lamela, C., Simal-Gándara, J., Falqué-López, E., 2013. Training of panelists for the sensory control of bottled natural mineral water in connection with water chemical properties. Food Chem. 141, 625–636.
Vreme, A., Pouligny, B., Nadal, F., Liger-Belair, G., 2015. Does shaking increase the pressure inside a bottle of champagne? J. Colloid Interface Sci. 439, 42–53.

```
# all code comes without any warranty
# the first section shows the code needed to do the analysis of covariance
# we imported a file named maya5wt.csv

maya5 <- read.csv(file = "maya5wt.csv", header = TRUE, sep = ",", dec =
".")
# we created a da.a data set as shown before in Case Studies 2 and 3
# as well as below
# do the ANCOVA with weight (wt) as part of the ANOVA
da.lm = lm(da.a ~ (wt + wine + rep + judge)^2 , data= da.d)
da.aov = aov(da.lm)
summary(da.aov)


# The next section shows the code for the
# MANOVA, ANOVA and CVA for RAW descriptive analysis data
# This code is identical to that showed in Case Study 2 except that we do
# not show lsd calculations here

library (agricolae)
library (candisc)
library (SensoMineR)

# we imported a file named maya5wt.csv which contained our raw descriptive
# analysis data
da.d <- read.csv(file = "maya5wt", header = TRUE, sep = ",", dec = ".")
dim(da.d)   # this showed us that the data was correct since 12*3*7 = 252
            # observations; 20 + 3 = 23 columns
head(da.d)
#  judge wine rep AllAroma AppleA CitrusA TropicalA StonefruitA ArtFruitA
#1    j1  8.1   1      2.6    2.1     3.2       0.0         0.0       0.0
#2    j2  8.1   1      3.5    0.0     0.0       0.0         0.0       2.5
#3    j3  8.1   1      2.3    0.0     0.0       0.0         1.1       0.0
#4    j4  8.1   1      7.2    2.2     5.9       7.7         8.9       7.6
#5    j5  8.1   1      2.3    0.0     0.0       1.0         0.0       0.0
#6    j6  8.1   1      0.0    4.8     2.2       0.0         1.6       0.0

# define judge, rep and wine as factors
# columns starting with a letter are automatically set as factors by R
# in our case judge and rep are not recognized as factors

is.factor(da.d$judge)        # is judge a factor?
#[1] TRUE                     # yes it is

is.factor(da.d$wine)         # is wine a factor?
#[1] FALSE                    # no it is not
da.d$wine = as.factor(da.d$wine)
is.factor(da.d$wine)
#[1] TRUE                     #now it is
is.factor(da.d$rep)          # is rep a factor?
da.d$rep = as.factor(da.d$rep)
is.factor(da.d$rep)          # now it is
```

```
# combine all attributes and define it as a matrix

da.a = as.matrix(da.d[,-c(1:3)])  # use all columns but the first
                                    three (wine, judge, rep)
head(da.a)
#     AllAroma AppleA CitrusA TropicalA StonefruitA ArtFruitA ConfectA
#[1,]      2.6    2.1     3.2       0.0         0.0       0.0      0.0
#[2,]      3.5    0.0     0.0       0.0         0.0       2.5      5.1
#[3,]      2.3    0.0     0.0       0.0         1.1       0.0      2.0
#[4,]      7.2    2.2     5.9       7.7         8.9       7.6      4.9
#[5,]      2.3    0.0     0.0       1.0         0.0       0.0      0.0
#[6,]      0.0    4.8     2.2       0.0         1.6       0.0      2.8

# create a 3-way MANOVA model with all 2-way interactions

da.lm = lm(da.a ~ (wine + rep + judge)^2 , data= da.d)

# run the MANOVA

da.maov = manova(da.lm)
summary(da.maov, test='Wilks') # print MANOVA table, see
#             Df   Wilks approx F num Df den Df    Pr(>F)
#wine          1 0.77675   3.2739     23  262.0 1.883e-06 ***
#rep           2 0.69970   2.2268     46  524.0 1.492e-05 ***
#judge        10 0.00047  13.7584    230 2436.3 < 2.2e-16 ***
#wine:rep      2 0.82741   1.1318     46  524.0    0.2613
#wine:judge   10 0.30108   1.4751    230 2436.3 1.163e-05 ***
#rep:judge    20 0.09379   1.4862    460 4038.4 8.818e-10 ***
#Residuals   284


# in this MANOVA wine, rep and judge and the interaction wine:judge and
# rep:judge are significant at p< 0.05.
# since wine was significant we continue with individual ANOVA's
# using the same lm model but now using the ANOVA output and lsds.
# See Case Study 2

da.aov = aov(da.lm)
summary(da.aov)  # print ANOVA tables for each attribute.


# continue further lsd, PCA, CVA analyses with only significant attributes
(significant # wine effect). See Table 3 for a means and lsds.

# To do the CVA
# create data subset with significant attributes only from original data
# sign. attributes are apple, citrus, confectionary, floral, yeast,
# nutty/oxidized, cooked apple, bitter, bubble size, bubble concentration


da.s = da.d[, c(2, 5, 6, 10, 11, 12, 13, 14, 23, 24, 25)]
head(da.s)
#  wine AppleA CitrusA ConfectA FloralA YeastBreadA NuttyOxidA CookAppleA
```

```
#1  8.1    2.1     3.2      0.0      0        0.0       0.0      0.0
#2  8.1    0.0     0.0      5.1      0        3.6       0.0      3.3
#3  8.1    0.0     0.0      2.0      0        1.5       1.5      1.8
#4  8.1    2.2     5.9      4.9      0        1.9       0.0      2.8
#5  8.1    0.0     0.0      0.0      0        2.0       3.0      0.0
#6  8.1    4.8     2.2      2.8      0        0.0       0.8      0.0

is.factor(da.s$wine)   # Wine is not a factor as its entries does not
start                                    with a letter
da.s$wine = as.factor(da.s$wine) #change to factor
is.factor(da.s$wine) #now it is a factor
# combine all attributes and define it as a matrix

da.as = as.matrix(da.s[,-1])
head(da.as)
#      AppleA CitrusA ConfectA FloralA YeastBreadA NuttyOxidA CookAppleA
#[1,]     2.1     3.2      0.0       0         0.0        0.0        0.0
#[2,]     0.0     0.0      5.1       0         3.6        0.0        3.3
#[3,]     0.0     0.0      2.0       0         1.5        1.5        1.8
#[4,]     2.2     5.9      4.9       0         1.9        0.0        2.8
#[5,]     0.0     0.0      0.0       0         2.0        3.0        0.0
#[6,]     4.8     2.2      2.8       0         0.0        0.8        0.0

library(candisc)

# build MANOVA model with significant attributes only, using only the wine
# effect. See Monrozier, R.; Danzart, M. (2001) Food Quality and
# Preference 12:393-406



da.s.lm = lm(da.as ~ wine, data = da.s) # run the one-way MANOVA

da.man1 = manova(da.s.lm)
summary(da.man1, test='Wilks') # print the one-way MANOVA table
#            Df   Wilks approx F num Df den Df    Pr(>F)
#wine         9 0.56083   2.0963     90 2119.6 1.779e-08 ***
#Residuals 320

da.cva = candisc(da.s.lm) # run the CVA
da.cva            # extract CVA output => eigenvalues etc.
#Canonical Discriminant Analysis for wine:
#       CanRsq Eigenvalue Difference   Percent Cumulative
#1 0.26453028 0.35967528    0.26461 56.190358     56.190
#2 0.08681043 0.09506288    0.26461 14.851221     71.042
#3 0.07019983 0.07549991    0.26461 11.794991     82.837

#Test of H0: The canonical correlations in the
#current row and all that follow are zero

#  LR test stat approx F num Df  den Df   Pr(> F)
#1      0.56083  2.34074     81 2025.28 4.468e-10 ***
#2      0.76255  1.36232     64 1811.83   0.03151 *
#3      0.83504  1.17926     49 1598.55   0.18715
```

```
# next we as to plot a CVA biplot, output not shown, used to get the
percentages explained, so that we could add that to the plots below.

plot(da.cva, type = 'n')

# For a better looking plots with 95% confidence intervals use the
# following code
# We will plot the first two canonical variates. If you want to plot
#canonical variables 1 and 3 then change da.cva$means[,1:2] to
#da.cva$means[, c(1,3] Figures 1 through 4

plot(da.cva$means[,1:2],xlab='CV1, 56.2.2%', ylab='CV2, 14.9%', cex=1.1,
xlim=c(-2,2), ylim=c(-2,2),axes=FALSE, cex.lab=.9)
#get the % variance ratio explained from the CVA output
axis(side = 1, at = c(-1.5,0,1.5), line=1, col='darkgray', cex.axis=.9)
axis(side = 2, at = c(-1.5,0,1.5), line=1, col='darkgray', cex.axis=.9)
abline(h=0,v=0, col='darkgray', lty=3, lwd=.8)
symbols(x=da.cva$means[,1], y=da.cva$means[,2],
circles=2/sqrt(table(da.cva$scores[,1])), add=TRUE, inches=FALSE, lty=2,
lwd=.5, fg='black')
text(da.cva$means[,1:2], row.names(da.cva$means), cex=0.6, pos=4,
col="blue")
plot(da.cva$structure[,1:2], pch='', xlab="CV1, 56.2", ylab="CV2, 14.9%",
xlim=c(-.6,.6), ylim=c(-.6,.6), axes=FALSE, cex.lab=.9)
#get the % variance ratio explained from the CVA output
Axis(side=1, at=c(-.6,0,.6), line=1, cex.axis=.9, col='gray')
Axis(side=2, at=c(-.6,0,.6), line=1, cex.axis=.9, col='gray')
arrows(0,0,da.cva$structure[,1],da.cva$structure[,2], length=0.1)
text(da.cva$structure[,1:2], labels=row.names(da.cva$structure),pos=4,
cex=.7)

# In this section we do use panellipse in the SensoMineR package on the
# RAW DA data. This code is very similar to the code in Case Study 3.
# We imported a file named Mead.csv
# 8 judges, 3 replicates, 14 wines, 24 sensory attributes
# We will only show some of the code for the MANOVA and ANOVAs, since it
# is very similar to the code shown in Case Study 2.
library (SensoMineR)


da.pcs = da.d[, c(1, 2, 5, 6, 10, 11, 12, 13, 14, 23, 24, 25)]
head(da.pcs)
#  judge wine AppleA CitrusA ConfectA FloralA YeastBreadA NuttyOxidA
#1    j1  8.1    2.1     3.2      0.0       0         0.0        0.0
#2    j2  8.1    0.0     0.0      5.1       0         3.6        0.0
#3    j3  8.1    0.0     0.0      2.0       0         1.5        1.5
#4    j4  8.1    2.2     5.9      4.9       0         1.9        0.0
#5    j5  8.1    0.0     0.0      0.0       0         2.0        3.0
#6    j6  8.1    4.8     2.2      2.8       0         0.0        0.8

panellipse(da.pcs, col.p = 2, col.j =1, firstvar =3, alpha = 0.05,
           coord = c(1,2), scale.unit = FALSE, nbsimul = 500,
           nbchoix = NULL, group = NULL, name.group = NULL,
```

```
        level.search.desc = 0.2, centerbypanelist = TRUE,
        scalebypanelist = FALSE, name.panelist = FALSE,
        variability.variable = TRUE, cex = 1, color = NULL)

# we perform the analysis on the significant attribute data frame. Our
# panelist information is in column 1 (col.j=1) and our wine information
# is in column 3 (col.p=3). Our first attribute is in column 3
#(firstvar=3) and we are calculating the first two principal components
# (coord = c(1,2). We want to use a covariance matrix (scale.unit = FALSE)
# and we are running 500 simulations to determine the 95% confidence
# ellipses. Additionally, the level.search.desc = 0.2 gives a threshold
above which an attribute is considered to NOT be discriminant according to
an ANOVA model: attribute=product + panelist. For our study this threshold
was reached with the attribute BITTER and thus the PCA output does NOT
have bitter in the multivariate space whereas the CVA output does.
```

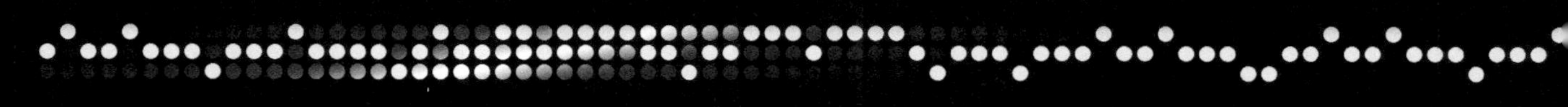

In this chapter

Objective

We were interested in characterizing the effects of multiple sip and salivary flow rates on the time course of astringency perception in a Cabernet Sauvignon wine.

Case Study 5

A time-intensity analysis of a Cabernet Sauvignon wine evaluated in multiple sips with and without added saliva and protein[a]

5.1 Panelist considerations

The effect of panelists' salivary flow rates and saliva composition on the perception of astringency in beverages has been extensively studied. There are different ways to determine salivary flow rates. The flow rates are determined either using a unilateral Carlson–Crittenden cap measuring parotid gland flow rates (Fischer et al., 1994) or whole mouth flow rates stimulated by citric acid and then expectorating into a cup (Ishikawa and Noble, 1995). Usually, the research shows that panelists with high salivary flow rates perceive lower intensities of astringency (Ishikawa and Noble, 1995) and have faster clearing rates of astringency (Fischer et al., 1994). However, Smith and Noble (1998) and Smith et al. (1996) found that there was no effect of salivary flow rates.

It is not only salivary flow rates but also saliva protein composition that affects the perception of astringency (Kallithraka et al., 2001). Dinella et al. (2010) used salivary protein concentration and saliva haze-forming capacity to define three panelist groups and showed that the high responding group, subjects whose salivary protein concentration decreased strongly after stimulation, found a 3-g/L tannic acid solution to be much more astringent than the low response or medium response groups.

From all the work done on saliva it is clear that it has an effect and we would suggest that when the focus of the study is purely on astringency, then the sensory scientist should take the time to measure panelist salivary flow rates, haze

[a]*Based on unpublished data collected by Inger Baek and Ann C. Noble. We thank these scientists for allowing us to use their data.*

Sensory and Instrumental Evaluation of Alcoholic Beverages. http://dx.doi.org/10.1016/B978-0-12-802727-1.00011-9

formation capabilities, and ideally protein fractions. On the other hand, if astringency is just one of the many attributes evaluated by the panel then we would suggest that this is not necessary, since we know, from Chapter 1, that our panelists will all have variable sensitivities to odors and tastes and thus focusing on the specific variability in astringency is not helpful.

5.2 Palate cleansers

It is known that repeated ingestion of astringent compounds leads to increases in perceived astringency; this may be especially true for panelists who have lower salivary flow rates (Noble, 1994; Courregelongue et al., 1999). These carry-over effects of the astringency can have serious implications in descriptive and analysis studies of astringent products, such as tannic red wines. Therefore, there has been an ongoing interest by sensory scientists to determine whether some palate cleansers work better than others when evaluating astringent samples. There have been a number of these studies, see for example, Vidal et al. (2016), Lee and Vickers (2010), Bajac and Pickering (2008), and Breslin et al. (1993). In these studies palate cleansers, such as water, crackers skim milk, carboxymethyl cellulose, wax and lemons, pectin solutions, plain sweetened yogurt, sucrose, artificial saliva, oil, and xanthan gum, etc. have been used. In all of these cases there was no consensus as to the ideal palate cleanser and none worked perfectly. As Vidal et al. (2016) said "none of the evaluated palate cleansers was efficient in preventing astringency buildup over repeated ingestions" and this highlights the importance of limiting the number of wine samples evaluated in a single session. They also suggest that a predetermined waiting time should be enforced between samples.

In our laboratories we enforce a 60-s wait between tannic red wine samples and provide the panelists with water and crackers (Nabisco Premium Saltine crackers, East Hanover, New Jersey). The panelists are instructed to rinse their mouths, eat a cracker, and rinse their mouths again. A timer on the computerized data entry system (FIZZ, Biosystèmes, Couternon, France) is also set to 60 s between evaluations. We also limit the number of samples to six, in rare cases eight.

5.3 Materials and methods

Twenty one panelists initiated the session using sensory software (FIZZ, Biosystèmes, Couternon, France), which prompted the subject to sip and spit each of four 10-mL Cabernet Sauvignon (Turning Leaf Cabernet Sauvignon, Modesto, California) samples. Panelists sipped the four samples at 25-s intervals, expectorating 8 s after each sip and rated intensity of astringency continuously.

Between samples, panelists rinsed with a palate cleanser (6 g/L gelatin for 15 s and distilled water) for 15 s, repeated this procedure, and then rested 30 s (these data were not collected in our laboratories and these authors chose to use a palate cleanser instead of our usual procedure of water, saltine crackers, and water with an enforced 60-s break between samples). Data were collected in triplicate.

Prior to the first TI session, panelists' individual flow-rate status was determined by weighing saliva elicited in 60 s by 4 g/L citric acid.

5.4 Data analyses

In this study the output data file from the TI study is a matrix with the panelists' individual salivary flow status, panelists' identification code, samples and replication as qualitative variables, and the perceived intensity of astringency at 1-s intervals, collected for 170 s. These data were averaged and then we used R to plot the TI curves. We also extracted TI parameters from the raw curves (Table CS5.1)

Table CS5.1 Time–intensity (TI) parameters extracted from TI curves for use in analyses of variance

TI parameter	Explanation
Imax1	Maximum intensity after first sip
Tmax1	Time to maximum intensity for first sip
Init2	Lowest intensity before second sip
TInit2	Time of lowest intensity before second sip
Imax2	Maximum intensity after second sip
Tmax2	Time to maximum intensity for second sip
Init3	Lowest intensity before third sip
TInit3	Time of lowest intensity before third sip
Imax3	Maximum intensity after third sip
Tmax3	Time to maximum intensity for third sip
Init4	Lowest intensity before fourth sip
TInit4	Time of lowest intensity before fourth sip
Imax4	Maximum intensity after fourth sip
Tmax4	Time to maximum intensity for fourth sip

and analyzed these in a one-way analysis of variance with panelists' flow-rate status as the main factor (R-codes not shown).

5.5 Results

Figs. CS5.1 and CS5.2 show the average curves for the three flow rates. Fig. CS5.1 shows the actual mean values for each time point (data were collected once a second for 170 s) for each flow-rate group. Fig. CS5.1 would usually not be published but we wanted to show the mean data prior to the spline fitting. Fig. CS5.2 shows the geometrically smoothed curves. From these curves it is clear that panelists with a high flow-rate status perceived the wine to be less astringent than panelists with medium or low flow rates. Additionally, for the high flow-rate panelists the perceived astringency did not increase over the four sips. For the medium and low flow-rate panelists the perceived intensity of astringency did increase over the four sips of Cabernet Sauvignon.

The mean TI parameters are shown in Table CS5.2. Please note that the values in Table CS5.2 do not exactly coincide with the values shown in Fig. CS5.2; this happens due to the spline smoothing of the curves and thus the values in Table CS5.2 are more accurate. The initial maximum astringency after sip

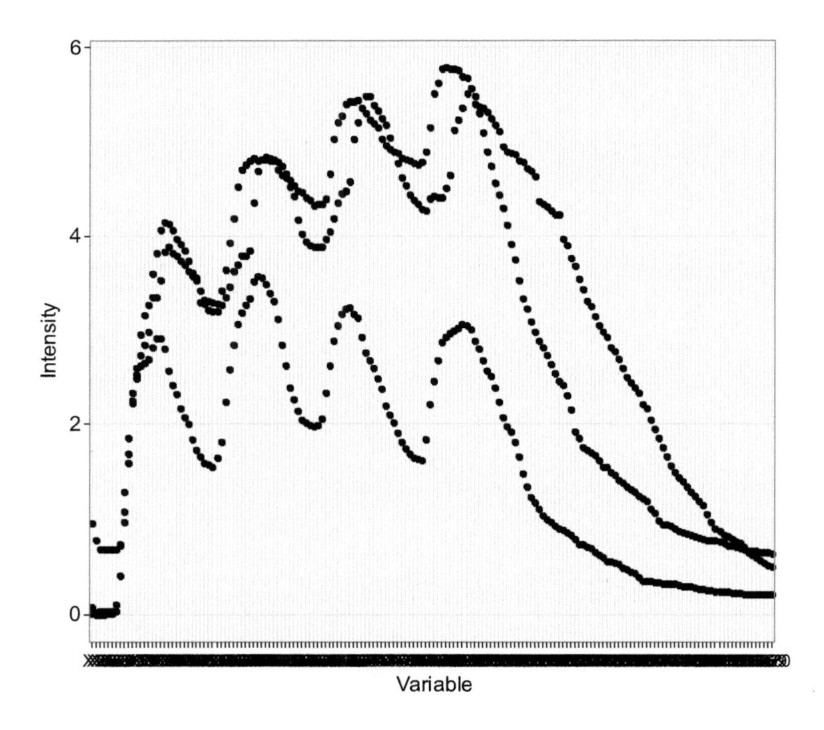

Figure CS5.1 Plot of mean time–intensity data points for panelist flow-rate status

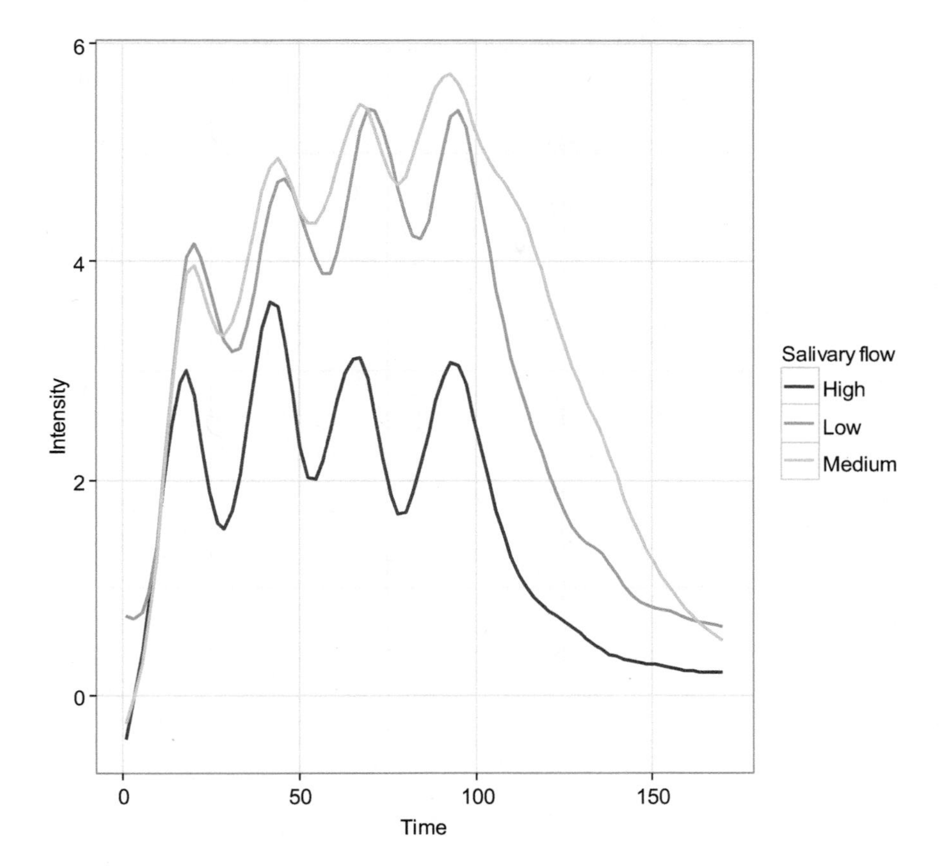

Figure CS5.2 Smoothed plot of mean time–intensity data for panelists flow-rate status Spline fitting used to smooth the curves.

1 is the same across flow-rate groups, however, for the maximum intensities after sips 2, 3, and 4 all clearly show that the high flow-rate group of panelists perceived the intensity of astringency a significantly lower than the medium and low flow-rate groups. Similarly the initial intensity prior to sips 2, 3, and 4 was significantly higher for panelists in the medium and low flow-rate groups than for those in the high flow-rate group. Finally, the high flow-rate group had a significantly shorter time to maximum astringency after sip 3 than the low flow-rate group.

5.6 Conclusions

The TI curves very clearly show the increase in the perceived astringency for the low and medium salivary flow group panelists. The high flow group does not perceive the astringency to increase over the four sips, probably due to the "washing out" effect of their saliva.

Table CS5.2 Means of TI parameters with Fisher's protected least significant difference for parameters differing significantly (21 panelists)

	TI parameters						
FRS	**Imax1**	**Tmax1**	**Imax2**	**Tmax2**	**Init2**	**TInit2**	**Imax3**
High	3.5	14.0	3.8 b	38.8	1.5 b	17.6	3.4 b
Medium	4.6	15.6	5.4 a	38.8	3.1 a	22.4	5.9 a
Low	4.5	15.4	5.2 a	38.2	3.3 a	22.5	5.8 a
lsd	—	—	1.30	—	1.31	—	1.24

	TI parameters						
FRS	**Tmax3**	**Init3**	**TInit3**	**Imax4**	**Tmax4**	**Init4**	**TInit4**
High	60.3 b	2.0 b	41.3	3.4 b	87.1	1.6 b	66.6
Medium	62.8 ab	4.3 a	41.6	6.6 a	89.8	4.7 a	65.2
Low	65.1 a	3.7 a	46.7	5.9 a	88.3	4.1 a	54.9
lsd	3.67	1.40	—	1.21	—	1.39	—

FRS, Flow-rate status, Imax1, maximum intensity after first sip; Imax2, maximum intensity after second sip; Imax3, maximum intensity after third sip; Imax4, maximum intensity after fourth sip; Init2, lowest intensity before second sip; Init3, lowest intensity before third sip; Init4, lowest intensity before fourth sip; lsd, Fisher's protected least significant difference; TI, time–intensity; TInit2, time of lowest intensity before second sip; TInit3, time of lowest intensity before third sip; TInit4, time of lowest intensity before fourth sip; Tmax1, time to maximum intensity after first sip; Tmax2, time to maximum intensity for second sip; Tmax3, time to maximum intensity after third sip; Tmax4, time to maximum intensity after fourth sip.
Means with the same letter do not differ significantly ($p < 0.05$)

References

Bajac, M.R., Pickering, G.J., 2008. Astringency: mechanisms and perception. Crit. Rev. Food Sci. Nutr. 48, 858–875.

Breslin, P.A.S., Gilmore, M.M., Beauchamp, G.K., Green, B.G., 1993. Psychophysical evidence that oral astringency is a tactile sensation. Chem. Senses 18, 405–417.

Courregelongue, S., Schlich, P., Noble, A.C., 1999. Using repeated ingestion to determine the effect of sweetness, viscosity and oiliness on temporal perception of soymilk astringency. Food Qual. Prefer. 10, 273–279.

Dinella, C., Recchia, A., Vincenzi, S., Tuorila, H., Monteleone, E., 2010. Temporary modification of salivary protein profile and individual responses to repeated phenolic astringent stimuli. Chem. Senses 35, 75–84.

Fischer, U., Boulton, R.B., Noble, A.C., 1994. Physiological factors contributing to the variability of sensory assessments: relationship between salivary flow rate and temporal perception of gustatory stimuli. Food Qual. Prefer. 5, 55–64.

Ishikawa, T., Noble, A.C., 1995. Temporal perception of astringency and sweetness in red wine. Food Qual. Prefer. 6, 27–33.

Kallithraka, S., Bakker, J., Clifford, M.N., Vallis, L., 2001. Correlations between saliva protein composition and some T–I parameters of astringency. Food Qual. Prefer. 12, 145–152.

Lee, C.A., Vickers, Z.M., 2010. Discrimination among astringent samples is affected by choice of palate cleanser. Food Qual. Prefer. 21, 93–99.

Noble, A.C., 1994. Application of time–intensity procedures for the evaluation of taste and mouthfeel. Am. J. Enol. Vitic. 46, 128–133.

Smith, A.K., June, H., Noble, A.C., 1996. Effects of viscosity on the bitterness and astringency of grape seed tannin. Food Qual. Prefer. 7, 161–166.

Smith, A.K., Noble, A.C., 1998. Effects of increased viscosity on the sourness and astringency of aluminum sulfate and citric acid. Food Qual. Prefer. 9, 139–144.

Vidal, L., Antunez, L., Gimenez, A., Ares, G., 2016. Evaluation of palate cleansers for astringency evaluation of red wines. J. Sens. Studies 31, 93–100.

```
# all code comes without any warranty
# we imported a file named Timean.csv
# 3 flow-rates, 170 variables
# plot of temporal TI data
# code with thanks to Scott Frost

library(reshape2)
library(ggplot2)
library(splines)
library(grid)

Tim=Timean
head(Tim)
dim(Tim)                            # 3 observation, 171 columns

# use melt
MtiAve <- melt(Tim)
head(MtiAve)                  #   salivaryflow variable      value
                              #   1         low      X1      0.96250000
                              #   2         high     X1      0.01111111
                              #   3       medium     X1      0.08095238

# plot the 170 observations for each salivary flow rate, see Figure 1
ggplot(MtiAve, aes(x=variable, y=value)) +
  geom_point() +
  labs(y = "Intensity")+      # this changes the y-axis label to intensity
  theme_bw()

# to color code by flow rate, add in the color call, figure not shown
ggplot(MtiAve, aes(x=variable, y=value)) +
  labs(y = "Intensity")+
  geom_point(aes(color=salivaryflow)) +
  theme_bw()


MtiAve$time <- rep(1:170, each=3) # this repeats each element 170 times
                                # and creates a new variable time to
                                # clean up the plot
head(MtiAve)                  # salivaryflow variable  value       time
                              # 1         low      X1  0.96250000    1
                              # 2        high      X1  0.01111111    1
                              # 3     medium       X1  0.08095238    1


# now we plot a black and white version, Figure 2

ggplot(MtiAve, aes(x=time, y=value)) +
  labs(y = "Intensity")+
  geom_smooth(aes(color=salivaryflow), method="lm", formula= y~ns(x,2  0),
se=FALSE) +
  scale_colour_grey()+
  theme_bw()



# and a colored version, figure not shown

ggplot(MtiAve, aes(x=time, y=value)) +
  labs(y = "Intensity")+
  geom_smooth(aes(color=salivaryflow), method="lm", formula= y~ns(x,2  0),
se=FALSE) +
  theme_bw()
```

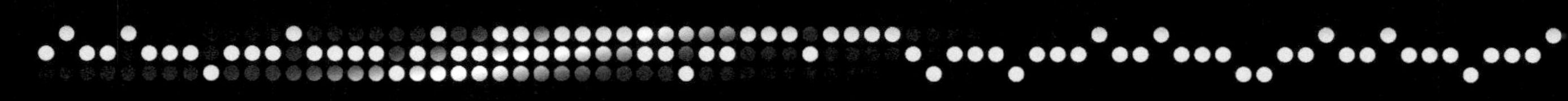

In this chapter

Objective

We were interested in characterizing the taste and mouthfeel TDS curves for three very different wines, a semisweet Riesling, a dry Rosé, and a Grenache–Syrah–Mourvedre Blend [These wines were a 2013 J. Lohr Monterey County White Riesling wine; a 2014 Charles & Charles Rosé wine (blend of Syrah, Mourvedre, Cabernet Sauvignon, Grenache, Cinsault, and Counoise; Columbia Valley, Washington); and a 2013 Tablas Creek Patelin de Tablas (blend of Grenache, Syrah, Mourvedre, and Counoise, Tablas Creek Vineyards, Paso Robles, California). The study was conducted in August 2015.].

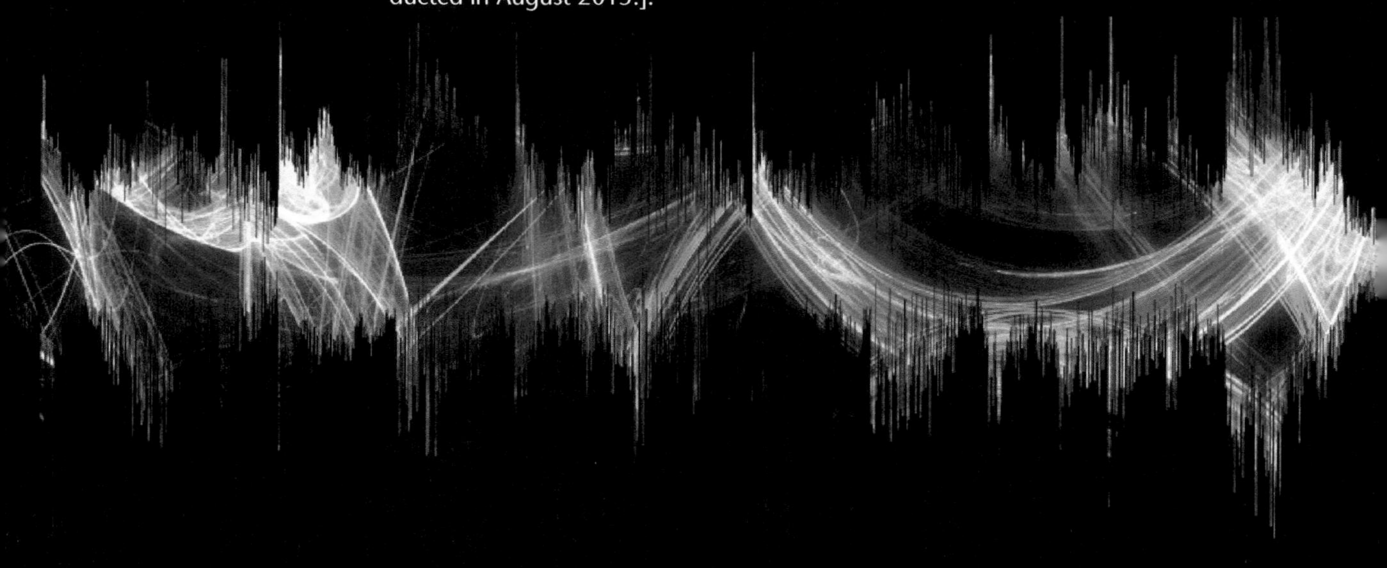

Case Study 6

A temporal dominance of sensation (TDS) analysis of three wines[a]

There are several panelist issues associated with temporal dominance of sensation (TDS) analysis that should be considered, see also Di Monaco et al. (2014): The panelists have to be trained to identify the sensations evaluated. This means that panelists should have reference standards available before each session and, similar to a descriptive analysis, panelists should be trained to agree on the concepts associated with each reference standard. Second, the panelists will be asked to indicate the dominant sensation over time and the dominance of a sensation may change slowly or very rapidly. Thus, the process is quite fatiguing and the panelists should not be asked to rate too many attributes (no more than six) within a sample or too many samples (depending on the product this may range from three for fatiguing samples—i.e., high alcohol and/or high astringency or sweetness, etc.,—to eight for fairly simple and less fatiguing samples) within a session. It is helpful to train panelists on the TDS data collection method by using the analogy of orchestral music to show the change in dominance for one or more musical instruments. We have used Benjamin Britten's "A Young Person's Guide to the Orchestra" (Ganeri et al., 1996) to good effect, although we did not use this music for this specific case study.

6.1 Materials and methods

Twenty-three panelists, all with prior experience in wine descriptive analysis, were trained to identify sweet, sour, and bitter tastes, as well as hot and astringent mouthfeel (Table CS6.1). Once trained to identify these sensory attributes the panelists were trained in the process of TDS. They received 20-mL cold coffee and were instructed to place the entire sample in their mouths, swish it around for 25 s, and then to expectorate the sample. A group discussion then followed of the progression of tastes that each panelist experienced. Each panelist was

---

[a] *Based on unpublished data collected by Scott Frost and Hildegarde Heymann.*

Sensory and Instrumental Evaluation of Alcoholic Beverages. http://dx.doi.org/10.1016/B978-0-12-802727-1.00012-0

Table CS6.1 Reference standards used for temporal dominance of sensation (TDS) analysis of bitter, sour, and sweet tastes and hot and astringent mouthfeel sensation in three wines

Reference standard	Recipe[a]
Bitter	800 mg/L caffeine (Sigma Aldrich, St. Louis, MO)
Sour	450 mg/L tartaric acid (Sigma Aldrich, St. Louis, MO)
Sweet	5 g/L sucrose (C&H Sugar, Crockett, CA)
Astringent	750 mg/L alum (Sigma Aldrich, St. Louis, MO)
Hot	7% ethanol (Everclear Grain Alcohol, 151 proof, Luxco, Inc. St. Louis, MO)

[a]*All solutions were made in Arrowhead Mountain Spring Water (Livermore, CA).*

then introduced to the FIZZ data collection terminal (Biosystèmes, Couternon, France), given an introduction to the tasting procedure, and allowed to practice with two wine samples.

During the data collection phase the panelist were provided with 20-mL samples and each sample for each panelist was given a unique, random three-digit code. Each panelist was presented in the booth, under red light, three wines in black ISO glasses, a red solo cup filled with water, a black solo cup for spitting, and two unsalted crackers. Panelists were instructed to match the serving order with the three-digit coded glasses. Once the correct wine was selected, they were instructed to take the entire 20 mL into their mouth and push the "start" button on the computer at the same time. The panelists then selected from the list on the computer screen the taste attribute which was the most dominant. This is a dynamic procedure, as the perception of which attribute is the most dominant changes with time. As the sensations changed the panelist changed the selected attribute. After 25 s, an on-screen message instructed the panelist to expectorate. The panelists continued to indicate dominant sensation(s) until the extinction of sensation or 120 s, whichever occurred first. Panelists were then asked to rinse their mouths with water and crackers while taking a 60-s break. Each sample was evaluated by each panelist in triplicate over three sessions.

6.2 Data analysis

The dominance rate at each time point (1–120 s) was calculated by summing the number of citations of each attribute at each time across all the panelists and then dividing by the number of observations (i.e., the total of the number of panelists times number of replications). Additionally, the TDS curves usually have a chance and significance line drawn on the curve. The chance level is defined as the number of attributes divided into 1. So, in our case with five attributes the chance level is 0.2. The significance rate is based on the binomial proportion of a normal approximation and it is the minimum dominance rate that is significantly higher than chance. The significance proportion is calculated using the following equation:

$$P_{\text{Significance}} = P_{\text{Chance}} + 1.645\sqrt{\frac{P_{\text{Chance}}(1 - P_{\text{Chance}})}{n}}$$

where n = number of observations (number of panelists times number of replications). For our study this would be 0.279.

We chose not to standardize the individual curves since we wanted to see the actual duration of the sensations. Depending on the study objective, standardizing curves may or may not be needed (Lenfant et al., 2009; Ng et al., 2012).

6.3 Results

Average TDS response curves for the three wines studied are shown in Fig. CS6.1. For the red wine (GSM) the perception of astringency dominated between 12.5 and 75 s. Additionally, bitterness also dominated the panelists' perceptions between 50 and 75 s. For the white wine (RIES) the sweetness dominated from

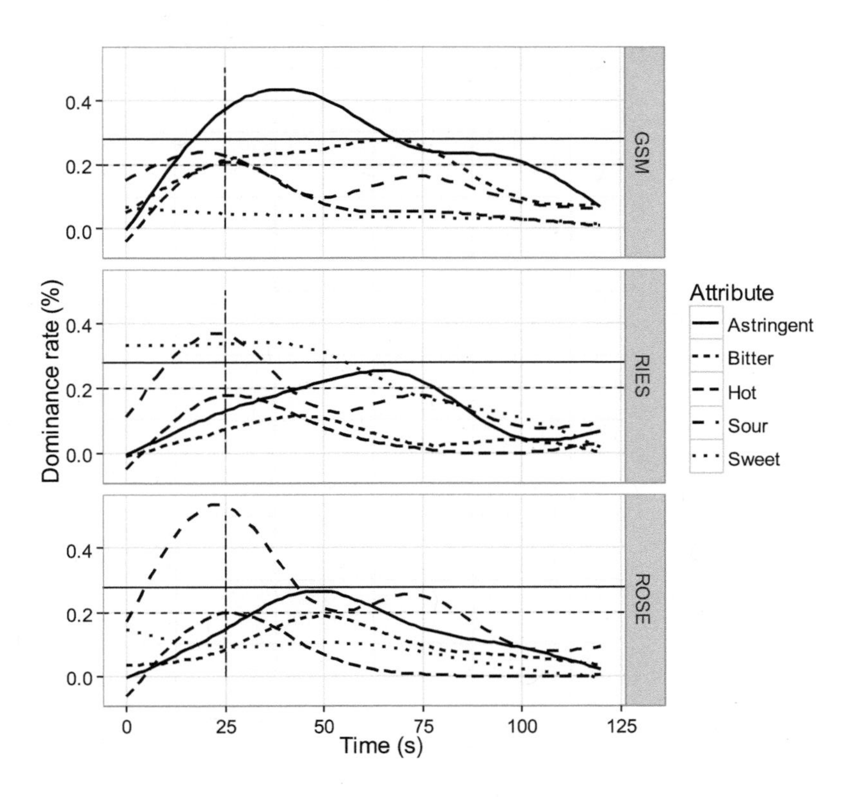

Figure CS6.1 TDS plots for three wines Time of expectoration indicated by *vertical line* at 25 s. Chance dominance rate indicated by *dashed horizontal line* and significant dominance rate indicated by *solid horizontal line*. GSM, Red wine from Tablas Creek Vineyard; RIES, white wine from J. Lohr Vineyards and Winery; ROSE, Rosé wine from Charles & Charles Winery.

the beginning to about 60 s with the perception of hotness also showing some dominance from about 10 to 38 s. Finally, for the rosé wine (ROSE) the dominant attribute was hotness and this lasted from about 5 to 45 s. There were a few panelists who also thought that astringency was slightly dominant at about 50 s and then hotness again became slightly dominant at about 70 s.

6.4 Conclusion

The TDS curves clearly showed very different dominant time courses for the taste and mouthfeel attributes in the three wines.

References

Di Monaco, R., Su, C., Masi, P., Cavella, S., 2014. Temporal dominance of sensations: a review. Trends Food Sci. Technol. 38, 104–112.

Ganeri, A., Britten, B., Kinsley, B., 1996. The Young Person's Guide to the Orchestra (Book and CD). Harcourt Children's Books, Houghton Mifflin Harcourt, San Diego, CA.

Lenfant, F., Loret, C., Pineau, N., Hartmann, C., Martin, N., 2009. Perception of food oral breakdown: the concept of sensory trajectory. Appetite 53, 659–667.

Ng, M., Lawlor, J.B., Chandra, S., Chaya, A., Hewson, L., Hort, J., 2012. Using quantitative descriptive analysis and temporal dominance of sensations analysis as complementary methods for profiling commercial blackcurrant squashes. Food Qual. Prefer. 25, 121–134.

```
# all code comes without any warranty
# we imported a file named tdsfinal.csv
# 3 ProductName (wines), 5 AttributeName (attributes), 120 seconds
# plot of TDS data
# code with thanks to Scott Frost


library(ggplot2)
library(grid)
library(gridExtra)
library(splines)
library(reshape2)

#import dataset tdsfinal and change name to TDSdf
TDSdf<-tdsfinal
str(TDSdf)  # Ensure ProductName,,AttributeName are factors
            # and time points are integer

#'data.frame': 1035 obs. of  123 variables:
# $ ProductName  : Factor w/ 3 levels "GSM","RIES","ROSE": 1 1 1 1 1 1 1 1
# 1 1 ...
# $ AttributeName: Factor w/ 5 levels "Astringent","Bitter",..: 5 4 2 1 3
# 5 4 2 1 3 ...
# $ X0           : int  0 0 0 0 0 0 0 0 0 0 ...
# $ X1           : int  0 0 0 0 0 0 0 0 0 0 ...

# sums the number of times each attribute was rated dominant for each
wine

tdsSum <- aggregate(TDSdf[,-c(1:2)], by=list(TDSdf$ProductName,
TDSdf$AttributeName), sum)

# calculate dominance rate by dividing sum for each attribute and wine
# combination by maximum possible sum or in this case 69
# 23 judges times 3 reps

tdsDOM <- aggregate(TDSdf[,-c(1:2)], by=list(TDSdf$ProductName,
TDSdf$AttributeName), FUN = function(i) sum(i)/69) head(tdsDOM)
head(tdsDOM)

# Group.1    Group.2         X0         X1         X2         X3  ....
# 1     GSM Astringent 0.00000000 0.00000000 0.00000000 0.00000000 ....
# 2    RIES Astringent 0.00000000 0.00000000 0.00000000 0.00000000 ....
# 3    ROSE Astringent 0.00000000 0.00000000 0.00000000 0.00000000 ....
# 4     GSM     Bitter 0.00000000 0.00000000 0.00000000 0.02898551 ....

# melt allows us to transpose the data set from short and fat to long
# and skinny
tdsMelt <- melt(tdsDOM, id.vars = 1:2)
# add the time points and change Group.2 to Attribute
tdsMelt$time <- as.numeric(gsub(x = tdsMelt$variable, pattern = "X",
replacement = ""))
tdsMelt$Attribute<-tdsMelt$Group.2
head(tdsMelt)

#  Group.1    Group.2 variable      value time  Attribute
# 1     GSM Astringent       X0 0.00000000    0 Astringent
# 2    RIES Astringent       X0 0.00000000    0 Astringent
```

```
# 3    ROSE Astringent       X0 0.00000000    0 Astringent
# 4     GSM     Bitter       X0 0.00000000    0     Bitter
# 5    RIES     Bitter       X0 0.01449275    0     Bitter
# 6    ROSE     Bitter       X0 0.00000000    0     Bitter


########  Graphics  ###########################
## TDS but use facet to plot by wine on same figure
## facet grid
ggplot(tdsMelt, aes(x=time, y=value)) +
geom_smooth(aes(linetype=Attribute), color="black", method="lm",
formula= y~ns(x,5), se=FALSE) +

# used 5 df for the spline smoothing.
# This was based on visual inspection of the plotted smoothed lines
# versus the actual data points.

# plots the vertical expectoration line
  geom_segment(aes(x = 25, y = 0, xend = 25, yend = .50), linetype=5)+
geom_abline(intercept=0.17, slope=0, linetype=2) +
  geom_abline(intercept=0.246, slope=0) +

#subsets into three plots, one for each wine
  facet_grid(Group.1 ~ .) +   ylab("Dominance Rate (%)") +
  xlab("Time (seconds)") +
  theme_bw()
```

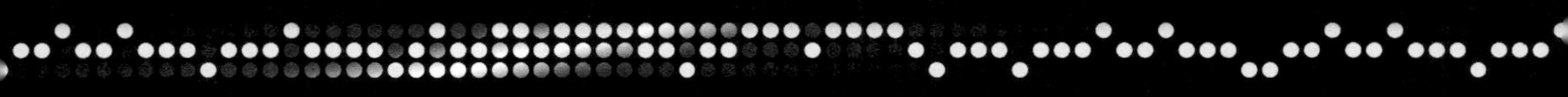

In this chapter

Objective

We were interested in characterizing the holistic differences among cream liqueurs.

Case Study 7

A sorting study using cream liqueurs[a]

Cream liqueurs (CLs) are alcoholic beverages containing dairy cream. CLs are traditionally made with Irish whiskey (e.g., Bailey's Irish Cream) but other alcohols are also used, such as rum and bourbon whiskey. Additionally, these beverages may be flavored with fruit, coffee, chocolate, etc. Bailey's Irish Cream is guaranteed to be shelf stable at 0–25°C for at least 6 months, after opening, but depending on the CL the shelf life could be as long as 2 years (Heffernan et al., 2011). Due to the stability of these products one does not have to worry about product deterioration during the course of an approximately 4–5-week sensory project. The result is that many fewer bottles of product are used than say with wines, where a new bottle has to be opened for each session.

7.1 Background

The data described here were collected as the first step in a study that eventually included a descriptive analysis (DA). The DA was performed on a subset of the CLs and that data are not used in this case study. A local alcoholic beverage retailer (Total Wine and More, Sacramento, California) supplied us with 13 CLs. The samples did not include fruit-, coffee-, or chocolate-flavored CLs, but rather focused on those that included the word *cream* on the label. For practical reasons, we could only use eight samples for the DA and consumer studies, and thus, we needed to find a way to choose a subset of samples that would be representative of the complete CL product space available in Northern California. The easiest way to do this was to have some consumers sort the CLs based on similarity and then analyze the data with multidimensional scaling (MDS) to determine which samples were very similar and which ones were very different from one another.

[a]*Based on unpublished data collected by Vanessa Rich and Hildegarde Heymann.*

Sensory and Instrumental Evaluation of Alcoholic Beverages. http://dx.doi.org/10.1016/B978-0-12-802727-1.00013-2

7.2 Materials and methods

Table CS7.1 shows the CLs used for this study, 1 of the 13 samples (Bailey's Original Irish Cream) was used as a blind duplicate, thus, the panelists received 14 samples. Seventeen panelists received 30 mL of each CL in clear ISO glasses labeled with three-digit codes. [We chose to use clear rather than black glasses because we wanted the panelists to also use the color of the samples as part of

Table CS7.1 Cream liqueurs (CLs) used for the sorting study

Code	Product	Alcohol type	ABV[a]	Producer	Cluster[b]
BAILEYS 2BAILEYS	Bailey's Original Irish Cream	Irish whiskey	17.0	R. & A. Bailey Co., Dublin, Ireland	1
BRENDANS	Saint Brendan's Irish Cream	Irish whiskey	17.0	Saint Brendan's Irish Cream Liqueur Co., Derry, Ireland	2
BROWNJUG	Brown Jug Bourbon Cream	Bourbon whiskey	17.0	A. Hardy Co., Des Plaines, IL, USA	3
CAROLANS	Carolan's Irish Cream	Irish whiskey	17.0	Gruppo Campari, Milan, Italy	2
CARUVA	Caruva Horchata Cream	Rum	12.5	United States Distilled Products (USDP), Princeton, MN, USA	4
CHESTNUT	Chestnut Ridge Bourbon Cream	Bourbon whiskey	17.0	USDP, Princeton, MN, USA	2
CHILA	Chila 'Orchata Rum Cream	Rum	13.75%	Chila 'Orchata, Louisville, KY, USA	4
EMMETS	Emmets Irish Cream	Irish whiskey	17.0	R & J Emmet and Co., Dublin, Ireland	2
FENNELLEYS	Fennelly's Irish Cream	Irish whiskey	17.0	Robert A. Merry Co., Clonmel, Ireland	1
KAVANAGH	Kavanagh Irish Cream	Irish whiskey	17.0	Kilbeggan Distillery, Kilbeggan, Ireland	2
OMARAS	Omara's Irish Cream Liqueur	Irish whiskey + wine	13.9	First Ireland Spirits, Abbeyleix, Ireland	1
OROVANA	Orovana Rum Cream Liqueur	Rum	17%	Destilerías Campeny, Barcelona, Spain	3
SHEELIN	Sheelin Irish Cream	Irish whiskey	13.9%	Unknown Irish Producer for Total Wines and More, Potomac, MD, USA	2

[a] *As listed on the label.*
[b] *Clusters were derived from the multidimensional scaling (MDS) plot.*

their sorting decisions. If we had wanted them to only use flavor (odor, taste, and mouthfeel) we would have used black glasses.] All serving orders were randomized across the panelists to prevent sample bias. Panelists were asked to sort the CLs into at least 2 and no more than 13 groups such that CLs sorted into the same group were more similar to one another than CLs sorted into different groups. Panelists were free to group as many products into one group as they wished, as long as in the end they had at least 2 groups and no more than 13 groups. Panelists entered their data into FIZZ (Biosystèmes, Couternon, France), and were provided with unsalted crackers and water to cleanse their palates as needed.

7.3 Data analysis

The data were entered into R and a nonmetric MDS analysis was performed on the data. As discussed by Lahne et al. (2016) either MDS or DISTATIS could be used for this type of data; we chose MDS for this case study. Cluster analyses (Wards, complete and average linkages) of the positions of the samples in the two-dimensional MDS map were performed to determine the number of clusters in the data. The x and y positions of the chosen clusters on the MDS map were evaluated by analysis of variance (ANOVA) and Fisher's protected least significant difference (lsd) was used to determine if these clusters were significantly different from one another ($p < 0.05$).

7.4 Results

The panelists created a median of 7 groups (range 4–9; mean 6.3 ± 1.6) from the 14 samples provided. The Kruskal Stress -1 value in percent for the three-dimensional solution was 12%. Based on Kruskal's (1964) discussion of stress, this falls in the fair category. Based on the Wards hierarchical clustering linkage dendrogram (Fig. CS7.1), we chose four clusters (Table CS7.1). We then

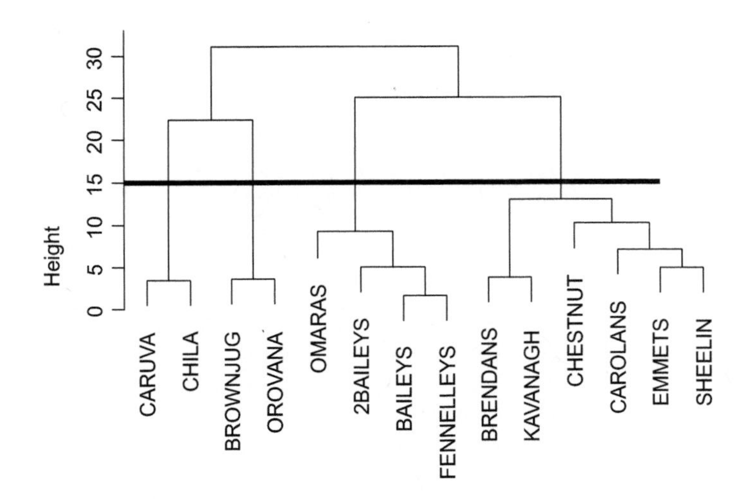

Figure CS7.1 Ward's hierarchical clustering linkage dendrogram for 13 CL samples The *horizontal black line* indicates the "cut" line for the four cluster solution. Bailey's Irish Cream was used as a blind replicate. For the identifying codes see Table CS7.1.

performed an ANOVA on the positions of the clustered CLs in dimensions 1, 2, and 3, respectively, of the MDS space and the clusters differed significantly in all dimensions. Table CS7.2 shows the ANOVAs and the mean cluster positions in the three-dimensional MDS space. All pairwise comparisons of clusters differ significantly in at least one dimension of the three-dimensional MDS space with the exception of clusters 2 and 4, and clusters 2 and 3, that differ significantly in all three dimensions (Table CS7.2). In cluster 4 the two Rum-based CLs with Horchata (in this situation Horchata is a riced-based Mexican beverage and it was used as the flavoring in these two CLs) (CARUVA and CHILA) were grouped together by the panelists, and also had the lowest alcohol by volume (ABV) content. The BROWN JUG, a Bourbon-based CL, and OROVANA, another Rum-based CL, were clustered together, both of these CLs had 17% ABV. The two BAILEYS samples were colocated and they were also clustered with FENNELLEY and OMARA; all of these except OMARA had 17% ABV and all, including OMARA, were Irish whiskey based. OMARA only had 13.9% ABV and was a blend of Irish whiskey and wine. Lastly all the other CLs formed a cluster. All of these, except

Table CS7.2 Analyses of variance (ANOVA) and means for clusters in the MDS map shown in Fig. CS7.1

Factor	df	Mean squares	*F*-value
Dimension 1			
Clusters	3	82.77	15.13 *
Residuals	10	5.471	
Dimension 2			
Clusters	3	59.546	6.77 *
Residuals	10	8.796	
Dimension 3			
Clusters	3	61.870	7.41 *
Residuals	10	8.346	
Cluster	**Dimension 1[a]**	**Dimension 2**	**Dimension 3**
1	4.0 a	−3.3 b	2.5 ab
2	1.5 a	−1.7 b	−1.2 b
3	−4.9 b	−3.9 b	−6.7 c
4	−7.7 b	−3.3 b	5.5 a

[a]*Means with the same letter do not differ significantly ($p < 0.05$). No least significant difference (lsd) values are shown because each pairwise comparison has a different lsd due to unequal numbers of CLs in each cluster.*
df, degrees of freedom; *, significance at $p < 0.05$

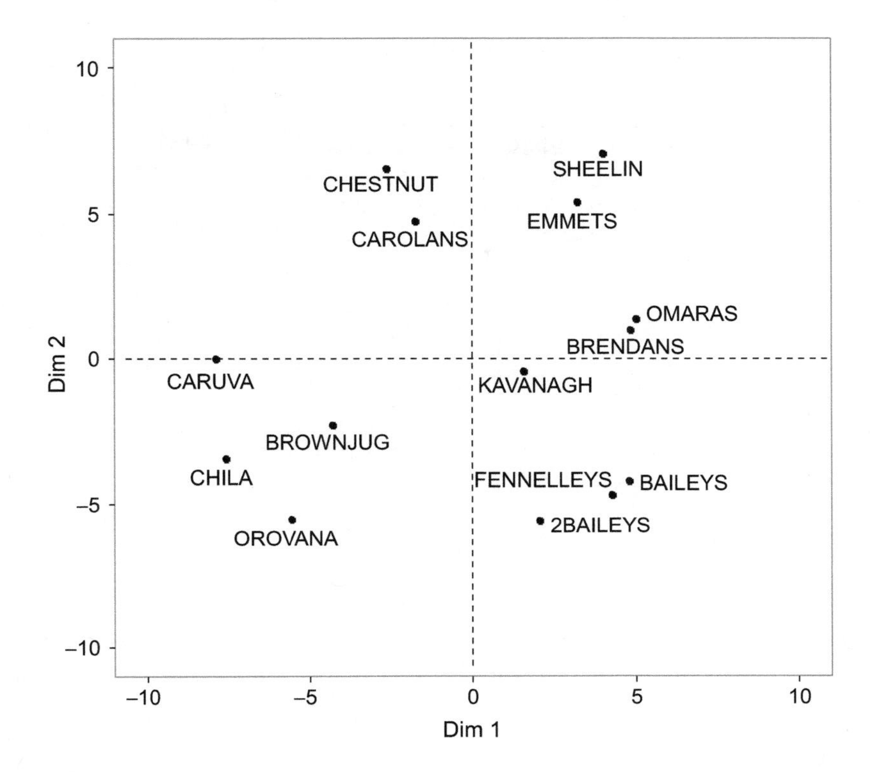

Figure CS7.2 **First two dimensions of the three-dimensional MDS plot of the CL samples** Bailey's Irish Cream was used as a blind replicate. For the identifying codes see Table CS7.1.

CHESTNUT, were Irish whiskey based and had 17% ABV, CHESTNUT had 17% ABV, but was Bourbon based.

In the two-dimensional plot of the MDS space (Fig. CS7.2) the OMARAS seems to cluster with group 2 rather than group 1 but in the third dimension (data not shown) it clusters with the other group 1 CLs. In this study, lower ABV CLs did not necessarily cluster together but the Horchata-flavored samples clustered differently than all the others. The two Bourbon-based CLs were very different from each other and did not cluster together. Lastly, the Irish whiskey–based CLs clustered in two different groups.

7.5 Conclusion

From this cluster and MDS analyses, we could choose the following eight products as representative of the product space for further DA and consumer analysis: CARUVA or CHILA, BROWNJUG, OROVA, OMARA, BAILEYS, CHESTNUT, BRENDANS or KAVANAGH, CAROLANS or EMMET or SHEELIN.

References

Heffernan, S.P., Kelly, A.L., Mulvihill, D.M., Lambrich, U., Schuchmann, H., 2011. Efficiency of a range of homogenization technologies in the emulsification and stabilization of cream liqueurs. Innov. Food Sci. Emerg. Technol. 12, 628–634.

Kruskal, J.B., 1964. Multidimensional scaling by optimizing goodness of fir to a nonmetric hypothesis. Psychometrika 9, 1–27.

Lahne, J., Collins, T.S., Heymann, H., 2016. Replication improves sorting-task results analyzed by DISATIS in a consumer study of American Bourbon and Rye Whiskeys. J. Food Sci. 81, S1263–S1271.

```
# all code comes without any warranty
# we imported a file named CLraw.csv
# the column headers were judge, product, group
# code with thanks to Pauline Lestringant

library(DistatisR)
library(dplyr)
library(reshape2)
library(MASS)
library(ggplot2)
library(agricolae)

# Import the data
d.indiv = read.csv(file = "CLraw.csv", header = TRUE, sep = ",", dec =
".")
d.indiv$Judge = as.factor(d.indiv$Judge)
str(d.indiv)

# 'data.frame': 238 obs. of  4 variables:
# $ Judge   : Factor w/ 17 levels "judge 1","judge 10",..: 1 1 1 1 1 1 ...
# $ Product: Factor w/ 14 levels "2BAILEYS","BAILEYS",..: 11 8 2 6 12 ...
# $ Group   : int  1 3 3 4 4 3 1 1 2 1 ...

# Create individual co-occurrence matrices
d.indiv.cast = acast(d.indiv, Product ~ Judge)
coo.matrix = DistanceFromSort(d.indiv.cast)

# Aggregate all co-occurence matrices
coo.matrix.aggr = apply(coo.matrix, MARGIN = c(1,2), FUN = sum)

# Run non-metric MDS on dissimilarity matrix for three dimensions
# K.stress is shown as % stress on the plot
K.stress = rep(0,3)
K.stress = data.frame(K.stress, row.names = paste("Dim", 1:3))
for (i in 1:3) {
  nm.mds = isoMDS(coo.matrix.aggr, k = i, trace = FALSE)
  K.stress[i, ] = nm.mds$stress
}
K.stress
# K.stress
# Dim 1 41.379550
# Dim 2 20.462239
# Dim 3 12.057548

plot(x = c(1:3), y = K.stress[,1], xlab = "Number of dimensions", ylab =
"Kruskall Stress")

# plot the MDS plot for dimensions 1 and 2
ggplot(data = data.frame(nm.mds$points), aes(x =
data.frame(nm.mds$points)[, 1], y = data.frame(nm.mds$points)[, 2])) +
  geom_text(aes(label = row.names(data.frame(nm.mds$points))), hjust=0.5,
vjust=1.5) +
  geom_point() +
  geom_hline(yintercept = 0, linetype = "dashed") +
  geom_vline(xintercept = 0, linetype = "dashed") +
  labs(x = "Dim 1", y = "Dim 2") +
  xlim(-10,10) +
  ylim(-10,10) +
  theme_bw() +
theme(panel.grid.major = element_blank(), panel.grid.minor =
element_blank())
```

```
# do hierarchical cluster analysis on the MDS dimensions
# we usually chose to evaluate the results from several linkage methods to
determine the underlying clusters
clus.ward = hclust(dist(nm.mds$points), method = "ward.D")
plot(clus.ward)
clus.complete = hclust(dist(nm.mds$points), method = "complete")
plot(clus.complete)
clus.average = hclust(dist(nm.mds$points), method = "average")
plot(clus.average)

# upon inspection of the dendrograms and our knowledge of the products,
# the four cluster Wards solution was the most logical to use.
# we then performed an ANOVA on the clusters for each MDS dimension

check.group = data.frame(nm.mds$points)
check.group$ward = as.factor(cutree(clus.ward, k = 4))
aov.ward = aov(nm.mds$points ~ ward, data = check.group)
summary(aov.ward)
#Response 1 :
#             Df  Sum Sq Mean Sq F value    Pr(>F)
#ward          3 248.316  82.772   15.13 0.0004784 ***
#Residuals    10  54.705   5.471
#Signif. codes:  0 '***' 0.001 '**' 0.01 '*' 0.05 '.' 0.1 ' ' 1
#
# Response 2 :
#             Df  Sum Sq Mean Sq F value   Pr(>F)
# ward         3 178.639  59.546  6.7696 0.009008 **
#Residuals    10  87.962   8.796
#Signif. codes:  0 '***' 0.001 '**' 0.01 '*' 0.05 '.' 0.1 ' ' 1
#
#Response 3 :
#             Df  Sum Sq Mean Sq F value  Pr(>F)
#ward          3 185.609  61.870  7.4132 0.00669 **
#Residuals    10  83.459   8.346
#Signif. codes:  0 '***' 0.001 '**' 0.01 '*' 0.05 '.' 0.1 ' ' 1


# since the ANOVA was significant for both dimensions we then performed an
# lsd for the four clusters on each MDS dimension
lsd.ward = lapply(1:ncol(nm.mds$points), function(i) {
aov.ward = aov(check.group[,i] ~ ward, data = check.group)
lsd = LSD.test(aov.ward, "ward")
lsd$groups
})
names(lsd.ward) = c("Dim 1", "Dim 2", "Dim 3")
lsd.ward
#$`Dim 1`
 #trt     means M
#1   1  4.040057 a
#2   2  1.536597 a
#3   3 -4.946074 b
#4   4 -7.743830 b

#$`Dim 2`
 #trt     means M
#1   2  4.064407 a
```

```
#2   4 -1.728605 b
#3   1 -3.275852 b
#4   3 -3.912911 b

#$`Dim 3`
 #trt     means  M
#1   4  5.528652  a
#2   1  2.474402 ab
#3   2 -1.248565  b
#4   3 -6.731761  c

# Now we plotted the two dimensional MDS plot indicating which cream
# liqueurs are in which clusters
ggplot(data = check.group, aes(x = check.group[, 1], y = check.group[, 2],
color = check.group[, 3])) +
  geom_text(aes(label = row.names(check.group)), hjust=0.5, vjust=1.5) +
  geom_point() +
  geom_hline(yintercept = 0, linetype = "dashed") +
  geom_vline(xintercept = 0, linetype = "dashed") +
  labs(x = "Dim 1", y = "Dim 2") +
  xlim(-10,10) +
  ylim(-10,10) +
  theme_bw() +
  theme(panel.grid.major = element_blank(), panel.grid.minor =
element_blank())
```

In this chapter

Objective

We were interested in using a rapid sensory technique, PM also known as napping, to characterize the holistic differences in the aromas of 12 gins.

Case Study 8

Projective mapping (PM) of gins from the United Kingdom, the United States of America, France, and Germany[a]

8.1 Materials and methods

Twelve gins were selected for this study (Table CS8.1). Eighteen panelists participated; some had previous experience in projective mapping (PM) and/or descriptive analysis though not on spirits. The 12 different products were presented to the panelists in each session and consisted of 25 mL aliquots in clear ISO glasses with a randomized three-digit code label. In this case the gins were served at full strength since they were only evaluated for aroma. However, we usually dilute all spirits 50:50 with water to decrease panelist fatigue. Each panelist was asked to place the products, using their own aroma attributes to differentiate among them, on a 45-cm^2 "square tablecloth." Finally they were asked to transfer their map to the FIZZ (Biosystèmes, Couternon, France) software, where the *X* and *Y* location coordinates of each gin were stored as a matrix for each product. Panelists performed the evaluation in duplicate, on consecutive days, without knowing that the products were identical for both sessions.

8.2 Data analysis

The data were analyzed using multifactor analysis (MFA) with each panelist and each replication as an individual data set. There were thus 36 (18 panelists times 2 replications) individual data sets. A bootstrap method was used to calculate 95% confidence ellipses around the mean position of each gin sample in the MFA space. Additionally, the RV coefficients of the individual panelist-replications were determined relative to the total MFA space.

---

[a] *Based on unpublished data collected by Jose Sanchez Gavito Sanchez and Hildegarde Heymann.*

Sensory and Instrumental Evaluation of Alcoholic Beverages. http://dx.doi.org/10.1016/B978-0-12-802727-1.00014-4

Table CS8.1 Brand name, producers, country of origin, and codes for gin samples used in this case study

Code	Brand name	Country	Producer
UK1	Bombay Original Dry	UK	Bacardi Co., Hamilton, Bermuda
UK2	Bombay Sapphire	UK	Bacardi Co., Hamilton, Bermuda
UK3	Booths	UK	Diageo PLC, Park Royal, UK
UK4	Tanqueray	UK	Diageo PLC, Park Royal, UK
UK5	Tanqueray 10	UK	Diageo PLC, Park Royal, UK
F1	Magellan Blue Gin	France	Angeac Distillery, Cognac, France
F2	Citadelle Gin	France	Maison Ferrand, Ars, France
G1	Doornkaat	Germany	Berentzen-Gruppe AG, Haselünne, Germany
G2	Schlichte	Germany	Schwarze und Schlichte Markenvertrieb GMBH & Co., Oelde, Germany
USA1	Cascade Mountain	USA	Bendistillery, Bend, OR, USA
USA2	New Amsterdam	USA	New Amsterdam Spirits, Modesto, CA, USA
USA3	No. 209	USA	Distillery No. 209, San Francisco, CA, USA

UK, United Kingdom; USA, United States of America.

8.3 Results

The average RV coefficient with the MFA solution was 0.50 ± 0.05 with all the panelists falling within a range of 0.60–0.39 (Table CS8.2). Low RV coefficients indicate that panelists do not show high agreement with the overall MFA product map. Based on other studies (Tomic et al., 2015) the RV coefficients found here

Table CS8.2 Thirty-six RV coefficients (18 panelists times 2 duplicates) of the panelists with multifactor analysis (MFA)

Group[a]	RV	Mean RV over reps
Pan1.1	0.600183	
Pan1.2	0.481885	0.541034
Pan2.1	0.511902	
Pan2.2	0.493474	0.502688
Pan3.1	0.530721	
Pan3.2	0.524376	0.527548
Pan4.1	0.503338	
Pan4.2	0.467141	0.485239

Table CS8.2 Thirty-six RV coefficients (18 panelists times 2 duplicates) of the panelists with multifactor analysis (MFA) (*cont.*)

Group	RV	Mean RV over reps
Pan5.1	0.570069	
Pan5.2	0.4645	0.517284
Pan6.1	0.530107	
Pan6.1	0.497195	0.513651
Pan7.1	0.45442	
Pan7.2	0.392777	0.423599
Pan8.1	0.58	
Pan8.2	0.599404	0.589702
Pan9.1	0.554187	
Pan9.2	0.518878	0.536533
Pan10.1	0.547337	
Pan10.2	0.571346	0.559341
Pan11.1	0.504519	
Pan11.2	0.501277	0.502898
Pan12.1	0.456503	
Pan12.2	0.499251	0.477877
Pan13.1	0.543107	
Pan13.2	0.462923	0.503015
Pan14.1	0.438995	
Pan14.2	0.423041	0.431018
Pan15.1	0.472011	
Pan15.2	0.503341	0.487676
Pan16.1	0.439706	
Pan16.2	0.443508	0.441607
Pan17.1	0.5256	
Pan17.2	0.526118	0.525859
Pan18.1	0.530666	
Pan18.2	0.47131	0.500988
Mean RV	0.503753	
Standard deviation	0.049497	

[a] *Group indicates panelists number followed by replication number.*

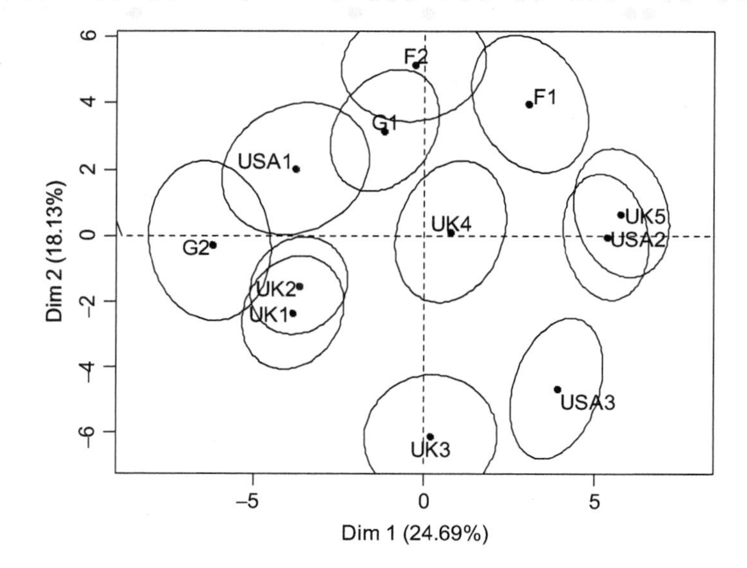

Figure CS8.1 MFA plot of 12 gin samples Score plot of dimensions 1 and 2. Please see Table CS8.1 for codes.

would indicate medium to low agreement, this is not very surprising since these were untrained panelists evaluating the aromas of a very challenging product.

The first three dimensions of the MFA captured 55.1% of the variance in the data set and thus we plotted the score plots for these three dimensions (we did not plot the loadings plots since they only showed the positions of the individual consumers and did not add to this discussion. Please see Case Study 10 for another use of MFA where we did plot the loadings plots). We could have also

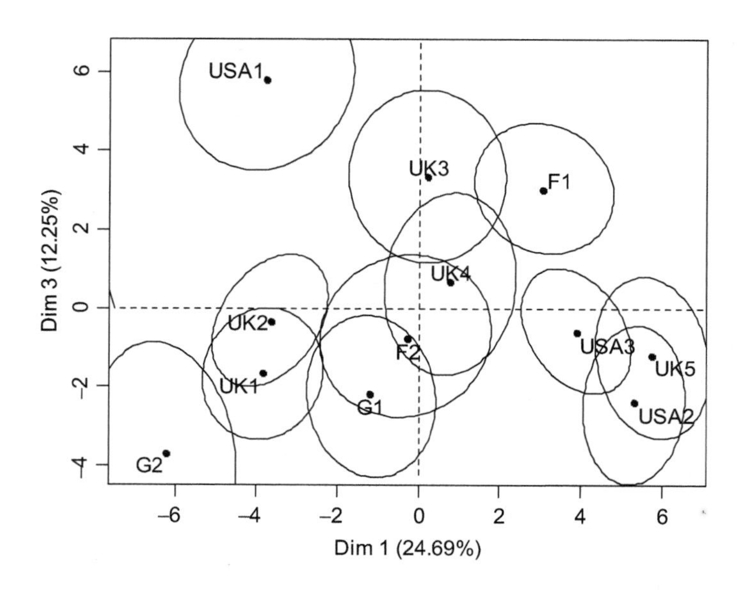

Figure CS8.2 MFA plot of 12 gin samples Score plot of dimensions 1 and 3. Please see Table CS8.1 for codes.

included the fourth dimension which would have added 9.5% to the total explained variance but chose not to in order to simplify this case study. Figs. CS8.1 and CS8.2 show the first and second, and first and third dimensions, respectively, of the MFA. Interestingly, the UK1 and UK2 samples were produced by the same company and did not differ significantly, whereas the UK5 and UK4 samples were also produced by the same, albeit different, company and these two brands were quite different from one another. UK5 and USA2 did not differ significantly from one another nor did UK1 and G2 or G1 and F2. All the other gin samples differed from one another in the three-dimensional MFA space as indicated by the nonoverlapping confidence ellipses.

8.4 Conclusions

The PM of the 12 gin samples showed that the majority of the brands were very different from one another. The exceptions were the two Bombay samples (UK1 and UK2), Doornkaat (G1), and Citadelle (F2) as well as Bombay Original Dry (UK1) and Schlichte (G2). Please keep in mind that these samples were only evaluated for aroma and not taste or mouthfeel.

Reference

Tomic, O., Berget, I., Naes, T., 2015. A comparison of generalized Procrustes analysis and multiple factor analysis for projective mapping data. Food Qual. Prefer. 43, 34–46.

```
# all code comes without any warranty
# we imported a file named gins.csv
# 3 ProductName (wines), 5 AttributeName (attributes), 120 seconds
# plot of TDS data
# code with thanks to Scott Frost

library(SensoMineR)

# Import the data
Gins <- read.csv(file = "gins.csv", header = TRUE, sep = ",", dec = ".")
head(gins)

#ProductName a1x a1y a2x a2y b1x b1y b2x b2y bok1x bok1y bok2x bok2y
#1          UK1  57  50  33  62  81  44  34  20    77    33    87    60
#2          UK2  43  53  37  56  33  51  63  75    23    22    90    76
#3          UK3   8   9   6  15   9   9  11  91    24    48    80    68
#4          UK4  77  88  14  82  40  49  35  15    19    54    12    53
#5          UK5  14  97  11  93   8  90  19  91    45    78    46    94
#6           F1  29  72  86  92  13  14  29  67    31    93    10    63

row.names(gins)=gins$ProductName
mfa.gins<-gins[,-1]
head(gins)

#     ProductName a1x a1y a2x a2y b1x b1y b2x b2y bok1x bok1y bok2x bok2y
#UK1          UK1  57  50  33  62  81  44  34  20    77    33    87    60
#UK2          UK2  43  53  37  56  33  51  63  75    23    22    90    76
#UK3          UK3   8   9   6  15   9   9  11  91    24    48    80    68
#UK4          UK4  77  88  14  82  40  49  35  15    19    54    12    53
#UK5          UK5  14  97  11  93   8  90  19  91    45    78    46    94
#F1            F1  29  72  86  92  13  14  29  67    31    93    10    63

# the following code performs the MFA as well as a bootstrap of the data
# to get the information for the confidence ellipses
mfa.all <- MFA(mfa.gins,group=rep(2,36), type= rep("c", 36),
name.group=paste("pan", 1:36), ncp=5)
prod= NULL
for(i in 1:nrow(gins)){prod=c(prod, rep(row.names(gins)[i], 36))}
ell.gins=simule(cbind(as.factor(prod), mfa.all$ind$coord.partiel), 500)
ell12 = coord.ellipse(ell.gins$simul, centre=NULL, axes=c(1,2),
level.conf=0.95)
ell13 = coord.ellipse(ell.gins$simul, centre=NULL, axes=c(1,3),
level.conf=0.95)

# this code shows the eigenvalues
mfa.all$eig

# eigenvalue percentage of variance cumulative percentage of variance
#comp 1   14.085649     24.694471              24.69447
#comp 2   10.342187     18.131563              42.82603
#comp 3    6.988973     12.252824              55.07886
#comp 4    5.440322     9.537784               64.61664
#comp 5    5.258491     9.219003               73.83565
#comp 6    3.904095     6.844524               80.68017
#comp 7    3.005634    5.269372                85.94954
#comp 8    2.653647    4.652281                90.60182
#comp 9    2.211052    3.876339                94.47816
```

```
#comp 10   1.842123    3.229545                97.70771
#comp 11   1.307517    2.292293                100.00000

# this code plots the MFA plots for dimensions 1 and 2 and dimensions 1
#and 3 - see Figures 1 and 2
plot.MFA(mfa.all, axes = c(1,2),choix="ind", habillage="none", ellipse=
ell12)
plot.MFA(mfa.all, axes = c(1,3),choix="ind", habillage="none", ellipse=
ell13)

# this code prints the RV coefficients, the contributions and the cos2
# values if needed by the sensory practitioner
mfa.all$group$RV
mfa.all$group$contrib
mfa.all$group$cos2
```

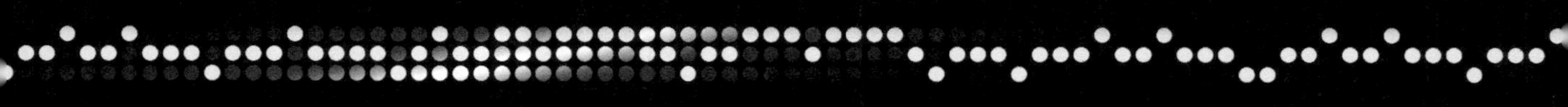

In this chapter

Objective

We were interested in evaluating consumer hedonic data for California Cabernet Sauvignon wines using ANOVA and internal preference mapping (IPM). Additionally, we used descriptive analysis (DA) data to interpret consumer liking data using external preference mapping (EPM).

Case Study 9

Internal (IPM) and external preference mapping (EPM) of Cabernet Sauvignon wines from California[a]

9.1 Materials and methods

The wines used in this study are described in Hopfer and Heymann (2013). For this case study we chose 12 of the original 27 different commercial California Cabernet Sauvignon wines (Table CS9.1). In Table CS9.1 and in all data outputs the original numbering for the complete set of 27 wines has been retained. This was done to make it easier for anyone interested in comparing the results of the subset to the entire published set of wines. The DA panel of 15 had 10 males (mean age of all panelists was 37 ± 17 years). Panelists underwent 6 1-h training sessions over a period of 2 weeks where they were exposed to subsets of the wines. During training, the panelists created, refined, and agreed upon sensory attributes (21 aromas, 3 tastes, 3 mouthfeel descriptors) that described the differences they perceived among the wines. For each sensory attribute, a corresponding reference standard was defined (Table CS9.2), and the panel had to blindly recognize these references at the end of the training, before they were able to proceed to the wine assessment. All wines were assessed in triplicate in individual tasting booths under red light and positive air pressure. During each session, the panelists were presented with six to seven wines in black tasting glasses (25 mL each), each labeled with an individual three-digit random code. Each attribute was rated on a computer screen on an unstructured line scale (0–9), anchored at both ends, using the FIZZ software (version 2.47B, Biosystèmes, Couternon, France).

One hundred seventy-four regular wine consumers participated in one session each. All consumers evaluated the wines the same day. Each consumer tasted six wines and rated their overall liking on an anchored, unstructured line scale

[a]*Adapted from data published data in Hopfer and Heymann (2013).*

Sensory and Instrumental Evaluation of Alcoholic Beverages. http://dx.doi.org/10.1016/B978-0-12-802727-1.00015-6

Table CS9.1 Some information about the 12 wines included in this case study

Code	Vintage	Ethanol (v/v %)	Closure	Retail price (per 750 mL)
W1	2008	14.3	Synthetic	$26.95
W5	2006	13.9	Natural	$15.00
W9	2009	14.5	Natural	$25.00
W10	2009	13.5	Natural	$9.99
W12	2009	13.5	Screw cap	$15.00
W15	2009	14.9	Natural	$24.99
W16	2011	13.5	Synthetic	$10.00
W18	2007	14.5	Natural	$70.00
W19	2010	13.5	Screw cap	$22.00
W20	2010	14.5	Natural	$19.99
W25	2008	14.7	Natural	$32.00
W26	2009	14.6	Natural	$59.00

Table CS9.2 Descriptive analysis reference standards for the significant sensory attributes

Aroma standards[a]		
Red fruit	One frozen strawberry (Dole, West Village, CA) + 5 frozen raspberries (Dole) in 15 mL base wine[b]	
Dark fruit	1 T Blueberry spread (Cascadian Farms, Rockport, WA) + 5 mL black cherry juice concentrate (RW Knudsen, Chico, CA) + 1 T blackberry jam (Mary Ellen, Orrville, OH) + 0.5 m black currant flavoring (IFF, New York, NY) + 3 mL water[c]	
Dried fruit	Dried fruit	One cut dried fig (SunMaid, Stockton, CA) + 10 raisins (SunMaid) + 1 cut dried apricot (SunMaid) in 15 mL base wine
	Oxidized	5 mL Marsala Superiore riserva 10 anni DOC (Marco de Bartoli) in 10 mL base wine
Fresh green	Herbal	0.05 g Dried dill (The Spice Hunter, San Luis Obispo, CA) + 0.1 g dried herb mix (Davis Co-Op, Davis, CA) in 30 mL base wine
	Grassy	Four fresh grass clippings
	Minty	Two crushed fresh mint leaves + 0.5 mL eucalyptus solution (3 drops eucalyptus essential oil in 100 mL water)

Table CS9.2 Descriptive analysis reference standards for the significant sensory attributes (*cont.*)

Canned veg	2 mL Canned asparagus brine (Green Giant, Minneapolis, MN) + 2 mL canned green bean brine (Green Giant) + 1 mL canned sweet corn brine (Best Yet, Keene, NH) in 15 mL base wine	
Oak	0.15 g Evoak Premium dark roasted chips (Oak Solutions, Napa, CA) in 15 mL base wine	
Sweet aroma	Honey	0.25 mL Vanilla extract (Kirkland, Costco) + 1 T Mrs. Richardson's Butterscotch caramel (Frankfort, IL)+ 1 T honey (Lienert's Mountain wildflower honey, Sacramento, CA)
	Chocolate	0.5 g Grated 70% chocolate (BRIX, Rutherford, CA) + 0.5 g 100% chocolate (Baker's, Kraft's Food, Northfield, IL) in 15 mL base wine
Spices	0.05 g Ground cloves (McCormick, Hunt Valley, MD) + 0.06 g ground nutmeg (McCormick) + 0.07 g ground ginger (McCormick) + 0.2 g ground cinnamon (McCormick) in 15 mL base wine	
Soy sauce	12 mL Soy sauce (Hisakawa, Golding Farms Foods, Winston-Salem, NC) in 15 mL base wine	
Brett[d]	20 mL Cabernet Sauvignon (Walton 2006, Napa Valley, CA) + 0.05 g white pepper (McCormick)	
Earthy	1/2 T Potting soil (Black Gold, Bellevue, WA) + 1 g Orchid bark (Black Gold) + 1/2 fresh button mushroom + 5 drops water	
Chemical	Verbal description: the smell of ammonia and chlorinated swimming pool	
Alcohol aroma	1 mL 95% Ethanol (Goldshield, Hayward, CA)	
Sulfur	Burnt rubber	1 mL 95% Ethanol (Goldshield, Hayward, CA) in 15 mL wine
	Rotten egg	1/2 Hard-boiled egg (boiled for 30 min) + 0.5 mL SO_2 solution
Taste and mouthfeel standards		
Sweet	10 g/L Sucrose (C + H, Crockett, CA)	
Bitter	0.8 g/L Caffeine (Sigma Aldrich, St. Louis, MO)	
Astringent	0.8 g/L Aluminum sulfate (McCormick)	
Hot	250 mL/L 40% Vodka	

[a]*All attributes were anchored with the words "low" and "high" at the end of the unstructured line scale.*
[b]*Franzia Vintner's Select Cabernet Sauvignon (Ripon, CA) was used as base wine.*
[c]*All standards requiring water were prepared in deionized water (Arrowhead, Nestle, Stamford, CT).*
[d]*Brett refers to the smell associated with a wine spoilage yeast,* Brettanomyces bruxellensis. *It is reminiscent of leather, sweaty horse, and barnyard.*

with the anchors "low" at the left end and "high" at the right end of the scale as provided by FIZZ. Consumers were encouraged to expectorate the sample. The presentation order was a balanced incomplete block design with carry-over control, using the algorithm of Wakeling and MacFie (1995).

9.2 Data analyses

All significance testing used an alpha level of 5%. Univariate analysis of variance (ANOVA) for each attribute using a three-way fixed effect model with all two-way interactions was performed (data not shown). Liking score values in the consumer data that were missing, due to the incomplete block design of the study, were imputed using a regularized iterative principal component algorithm as proposed by (Josse and Husson, 2012). Once a complete data set was obtained, all following data analyses used the imputed data set. Hedonic liking data from the consumer clusters were analyzed by a two-way ANOVA fixed effects model followed by posthoc analysis according to Tukey [honestly significant difference (HSD)]. Principal component analysis (PCA) was used to create the IPM space. Consumer segmentation based on IPM vectors was obtained by hierarchical clustering of consumer IPM vectors, using Euclidean distances and Ward's linkage. The resulting clusters were chosen visually where a large drop in the hierarchical tree height was observed. Additionally, PCA using mean DA data was used to create the EPM space. The IPM clusters were projected into this space.

9.3 Results

There were significant differences in consumer liking for the wines (Table CS9.3). On average, the least liked wine was W26, at $59 per 750 mL bottle (Table CS9.4). At the top of the hedonic range were 8 wines (W15, W1, W5, W9, W10, W12, W19, and W25), ranging in price from $9.99 to $32 per 750 mL bottle. However, mean hedonic ratings tend to obscure the fact that there were probably a wide range of consumer hedonic scores for each wine. Creating an IPM based on individual hedonic scores for all the consumers will allow us to visualize this variation. Fig. CS9.1A shows the score plot for the IPM of the wines. The wines are fairly evenly distributed across the four quadrants of the plot,

Table CS9.3 Analysis of variance for consumer hedonic data

Source of variation	Degrees of freedom	Mean squares	*F*-value
Panelists	173	29.922	7.38*
Wines	11	12.762	3.15*
Error	1903	4.056	

* Indicates significance at $p < 0.05$.

Table CS9.4 Mean hedonic values for wines[a] with Fisher's protected least significant difference (lsd) value

Wine	Mean hedonic rating[b]
W1	4.6 abcd
W5	4.6 abc
W9	4.6 abc
W10	4.7 abc
W12	4.7 abc
W15	5.0 a
W16	4.5 bcd
W18	4.4 cde
W19	4.9 ab
W20	4.2 de
W25	4.6 abd
W26	4.1 e
HSD	0.42

[a] *Wine codes shown in Table CS9.1.*
[b] *Letters indicate which wines differed significantly.*

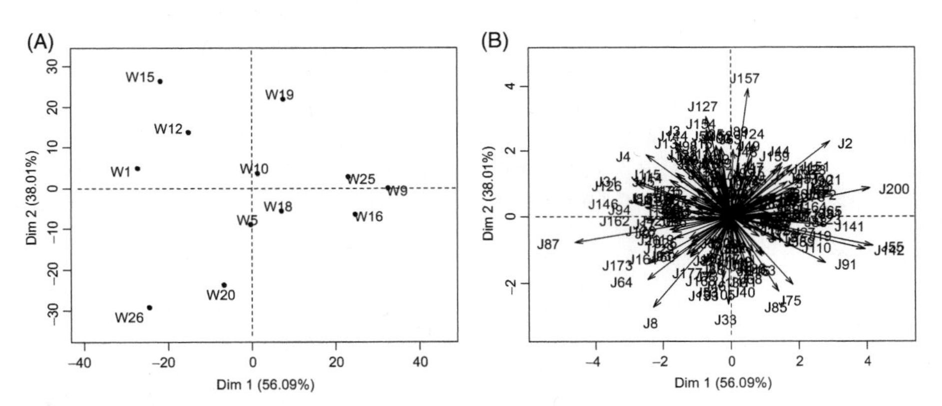

Figure CS9.1
(A) Internal preference map (IMP) score plot for 12 California Cabernet Sauvignon wines hedonically evaluated by 174 consumers. (B) IMP loadings plot for 12 California Cabernet Sauvignon wines hedonically evaluated by 174 consumers.

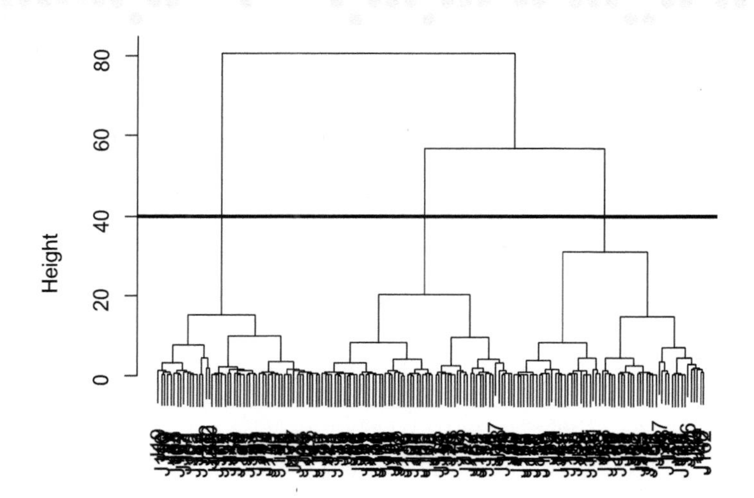

Figure CS9.2 **Dendrogram of Ward's hierarchical clustering of the 174 consumers in the internal preference mapping (IPM) space** The *horizontal black line* indicates the number of clusters chosen after visual inspection of the dendrogram.

additionally, the two dimensional IPM solution captures 94% of the variance in the data set. Fig. CS9.1B shows the loadings plots, where the vectors indicate the direction of preference for each of the 174 consumers. From this plot it is clear that the consumers were not in agreement about their degree of liking for each wine and every wine has consumers that liked it a great deal and other consumers that disliked it. This plot clearly shows that the mean values in Table CS9.4 obscure the actual variation in consumer liking for the wines and should be taken with a grain of salt. This type of distribution of consumer liking (where there are likers and dislikers of each product) is not unusual when one studies wines that are commercially available.

We wanted to see whether there were some clusters of consumers that were in agreement in terms of their likes and dislikes and thus, we did a Ward's hierarchical clustering on the vector positions for each consumer in the IPM. Based on visual inspection we chose a three cluster solution with the consumers fairly evenly split across the three clusters with 61, 52, and 61 consumers in clusters 1, 2, and 3, respectively (Fig. CS9.2). A one-way ANOVA showed that the clusters differed significantly in the two-dimensional IPM space (Table CS9.5 and Table CS9.6).

The external preference map (EPM) is a space based on the descriptive analysis (DA) data. For this we used the means of the significant attributes associated with the 12 wines. We had determined which of the attributes were significantly different using the methods described in Case Study 2 (data not shown). Of the 26 attributes used in the study, 19 were significantly different across the 12 wines used in this case study. For the EPM we used the means for each wine for each IPM cluster as the input on the consumer side. The results of the analysis is shown in Figs. CS9.3 and CS9.4, the three-dimensional DA derived space captured 75%

Table CS9.5 Analysis of variance of three cluster solution for dimension 1 and 2 of the IPM space

Source of variation	Degrees of freedom	Mean squares	*F*-value
Dimension 1			
Ward	2	96.614	94.9*
Error	171	1.018	
Dimension 2			
Ward	2	64.018	99.7*
Error	171	0.642	

* indicates significance at $p < 0.05$.

Table CS9.6 Mean values for each cluster

Cluster	Mean[a]
Dimension 1	
1	1.64 a
2	−0.56 b
3	−0.76 b
Dimension 2	
1	1.14
2	0.53
3	−0.87

[a]*Letters indicate which wines differed significantly.*

of the variance in the data set. The score plots shows that in dimensions 1 and 2 the wines were relatively evenly distributed across the dimension and in dimension 3 wine W1 was the driver of the space. Based on both the scores and loadings plots, wines W15 and W25 were sweeter, oaky, with dried and dark fruit aromas. On the other hand, wines on the left hand side of the plot were more Bretty, sulfury, and chemically, and wines in the north/top of the plot were more astringent. It seems that the consumers in cluster 1 liked sweeter wines with a hotter mouthfeel, while those in cluster 2 did not like astringent wines or wine W1. The consumers in cluster 3 liked astringent wines with vegetative and green notes.

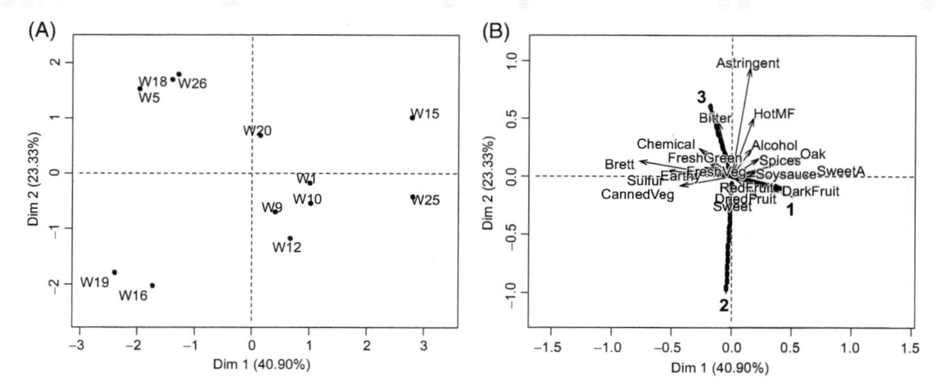

Figure CS9.3

(A) External preference map (EMP) score plot of dimensions 1 and 2 based on the means for the significant descriptive analysis (DA) attributes for 12 California Cabernet Sauvignon wines. (B) EMP loadings plot of dimensions 1 and 2 based on the means for the significant DA attributes for 12 California Cabernet Sauvignon wines. The position of the three clusters derived from the internal preference map (IMP) are shown in *black thick lines* and *bold font.*

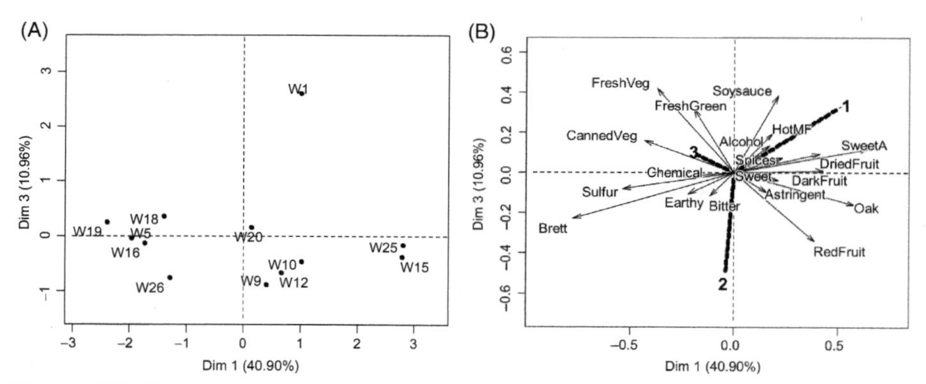

Figure CS9.4

(A) EMP score plot of dimensions 1 and 3 based on the means for the significant DA attributes for 12 California Cabernet Sauvignon wines. (B) EMP loadings plot of dimensions 1 and 3 based on the means for the significant DA attributes for 12 California Cabernet Sauvignon wines. The position of the three clusters derived from the IMP are shown in *black thick lines* and *bold font.*

9.4 Conclusions

In this case study we showed how to analyze consumer data using ANOVA, IPM, clustering, and EPM. We also showed that the mean liking scores often cover-up some very interesting consumer groupings as seen in the IPM clusters. Additionally, the use of EPM to determine drivers of liking is very useful in product development and blending operations.

References

Hopfer, H., Heymann, H., 2013. Judging wine quality: do we need experts, consumers or trained panelists? Food Qual. Prefer. 32, 221–233.

Josse, J., Husson, F., 2012. Handling missing values in exploratory multivariate data analysis methods. Journal de la Societe Francaise de Statistique 153, 79–99.

Wakeling, I.N., MAcFie, H.J.H., 1995. Designing consumer trials balanced for first and higher orders of carry-over when only a subset of k samples of t may be tested. Food Qual. Prefer. 6, 299–308.

```
# allcode comes without any warranty
# analysis of variance for consumer hedonic scores
# we imported a file named cons12.csv
# 12 wines 173 consumers

library(agricolae)
library(reshape2)
library(SensoMineR)
#Import the data
cons <- read.csv(file = "cons12.csv", header = TRUE, sep = ",", dec = ".")
head(cons)
#     wine      J10      J100      J101      J102      J103       J104      J105
# 1    W1 5.853851 6.800000 1.700000 0.6535121 4.158868 4.9000000 5.502061
# 2   W10 5.645779 5.529788 2.936924 3.1992017 3.787925 2.6491755 5.281937
# 3   W12 6.217736 6.292733 2.245496 2.2622614 5.032193 3.9723735 4.400000
# 4   W15 6.898470 7.052910 2.136123 2.1547734 8.200000 4.7000000 2.866272
# 5   W16 5.013484 5.800000 4.246275 4.3000000 2.440914 0.7350065 6.166438
# 6   W18 5.206546 4.800000 3.133009 3.4458000 2.000000 2.7000000 6.500000

#The file needs to be reformatted from 'short and fat' to 'long and
#skinny'
consMelt <- melt(cons, id.vars = 1)
head(consMelt)
#  wine variable    value
#1   W1      J10 5.853851
#2  W10      J10 5.645779
#3  W12      J10 6.217736
#4  W15      J10 6.898470
#5  W16      J10 5.013484
#6  W18      J10 5.206546

consMelt$judge<-consMelt$variable
head(consMelt)
#  wine variable    value judge
#1   W1      J10 5.853851   J10
#2  W10      J10 5.645779   J10
#3  W12      J10 6.217736   J10
#4  W15      J10 6.898470   J10
#5  W16      J10 5.013484   J10
#6  W18      J10 5.206546   J10

#check to see if both main effects are factors and they both were
is.factor(consMelt$wine)
is.factor(consMelt$judge)

#create matrix of dependent variables
da.a=as.matrix(consMelt [,-c(1:2, 4)])

#Now we want to do a 2-way ANOVA with no 2-way interactions and we want to
#see the output from the analysis
da.lm=lm(da.a~(judge + wine), data=consMelt)
da.aov=aov(da.lm)
summary(da.aov)
#               Df Sum Sq Mean Sq F value   Pr(>F)
#judge         173   5176  29.922   7.376  < 2e-16 ***
#wine           11    140  12.762   3.146 0.000314 ***
#Residuals    1903   7719   4.056

#now we ran an lsd on the means for the wines
```

```
lsd =LSD.test(lm(da.a~(judge+wine), data=consMelt), "wine")
lsd$groups
#   trt    means    M
#1  W15 4.971676    a
#2  W19 4.892966   ab
#3  W12 4.698231  abc
#4  W10 4.676615  abc
#5   W9 4.644682  abc
#6  W25 4.596715  abc
#7   W1 4.569648 abcd
#8  W16 4.498753  bcd
#9  W18 4.383378  cde
#10  W5 4.330438  cde
#11 W20 4.162579   de
#12 W26 4.066170    e
```

```
# to do the external preference map we imported the same consumer file as
# before
# we also imported a file named DA12.csv
# 12 wines 19 sensory attributes
# code with thanks to Pauline Lestringant


library(SensoMineR)
library(agricolae)
library(reshape2)

#read both tables WITH row.names
## Import the data
cons <- read.csv(file = "cons12.csv", header = TRUE, sep = ",", dec = ".")
head(cons[,1:5])
#  wine    J10     J100     J101      J102
#1 W1  5.853851 6.800000 1.700000 0.6535121
#2 W10 5.645779 5.529788 2.936924 3.1992017
#3 W12 6.217736 6.292733 2.245496 2.2622614
#4 W15 6.898470 7.052910 2.136123 2.1547734
#5 W16 5.013484 5.800000 4.246275 4.3000000
#6 W18 5.206546 4.800000 3.133009 3.4458000

#we want the row names to be the wines
row.names(cons)=gins$wine
cons<-cons[,-1]
head(cons[,1:5])
#         J10     J100     J101      J102     J103
#W1  5.853851 6.800000 1.700000 0.6535121 4.158868
#W10 5.645779 5.529788 2.936924 3.1992017 3.787925
#W12 6.217736 6.292733 2.245496 2.2622614 5.032193
#W15 6.898470 7.052910 2.136123 2.1547734 8.200000
#W16 5.013484 5.800000 4.246275 4.3000000 2.440914
#W18 5.206546 4.800000 3.133009 3.4458000 2.000000


DAmean <- read.csv(file = "DA12.csv", header = TRUE, sep = ",", dec = ".")
row.names(DAmean)=DAmean$wine
DAmean<-DAmean[,-1]
head(DAmean)
#    Alcohol    Brett CannedVeg  Chemical DarkFruit DriedFruit    Earthy
#W1  4.268889 1.642222 2.2466667 1.1222222  3.284444   2.855556 0.9222222
#W10 3.608889 2.020000 1.8866667 0.6955556  2.931111   2.913333 1.2377778
#W12 3.373333 2.177778 1.6777778 1.3733333  3.455556   2.382222 0.7755556
#W15 3.502222 1.737778 0.9955556 1.7533333  3.653333   2.382222 1.2800000
#W16 2.951111 3.793333 2.1688889 1.2155556  2.926667   1.808889 1.5866667
#W18 3.980000 3.793333 2.1066667 2.0600000  2.822222   1.986667 1.4355556

# Now we do an internal preference map using the consumers as variables
# This is a covariance PCA
ipm.pca = PCA(cons, scale.unit=FALSE, ncp=5, graph=TRUE)
ipm.pca # see Figure 1a and 1b

# do cluster analysis on the individual consumer IPM dimensions
# as shown in case study 8 we usually choose to evaluate the results from
# several linkage methods to determine the underlying clusters but here we
# only did Wards
```

```
clus.ward = hclust(dist(ipm.pca$var$coord), method = "ward.D")
plot(clus.ward) # see Figure 2


# upon inspection of the dendrogram
# the three cluster Wards solution was the most logical to use.
# we then performed an ANOVA on the clusters

check.group = data.frame(ipm.pca$var$coord)
check.group$ward = as.factor(cutree(clus.ward, k = 3))
check.group$ward
head(check.group)
#           Dim.1     Dim.2        Dim.3       Dim.4       Dim.5 ward
#J10   -0.07518513 0.7216310 -0.001689272 -0.30217726 -0.01715321    1
#J100  -0.34370927 0.6898849  0.016089606 -0.00957954 -0.04972099    1
#J101   1.05066584 0.4953184 -0.038838522  0.05567529  0.05074390    2
#J102   1.86140549 0.8732218  0.009623825 -0.07303648  0.31199509    2
#J103  -0.34787375 2.0445144  0.082396823 -0.05155367 -0.14891705    1
#J104  -1.19962290 0.5834600 -0.110441087  0.03435621  0.18817012    1

aov.ward = aov(ipm.pca$var$coord  ~ ward, data=check.group)
summary(aov.ward)
Response Dim.1 :
             Df Sum Sq Mean Sq F value    Pr(>F)
ward          2 193.23  96.614  94.934 < 2.2e-16 ***
Residuals   171 174.03   1.018

 Response Dim.2 :
             Df Sum Sq Mean Sq F value    Pr(>F)
ward          2 128.04  64.018  99.731 < 2.2e-16 ***
Residuals   171 109.77   0.642

 Response Dim.3 :
             Df Sum Sq  Mean Sq F value Pr(>F)
ward          2 0.0373 0.018671  0.4112 0.6635
Residuals   171 7.7643 0.045405

 Response Dim.4 :
             Df Sum Sq  Mean Sq F value Pr(>F)
ward          2 0.0891 0.044544  1.1893 0.3069
Residuals   171 6.4045 0.037454

 Response Dim.5 :
             Df Sum Sq  Mean Sq F value Pr(>F)
ward          2 0.0142 0.007082  0.2052 0.8147
Residuals   171 5.9016 0.034512

# since the ANOVA was significant for the first two dimensions we then
# performed an HSD test for the four clusters on each MDS dimension
hsd.ward = lapply(1:ncol(ipm.pca$var$coord), function(i) {
  aov.ward = aov(check.group[,i] ~ ward, data = check.group)
  hsd = HSD.test(aov.ward, "ward")
  hsd$groups
})
names(hsd.ward) = c("Dim 1", "Dim 2", "Dim 3","Dim 4","Dim 5")
hsd.ward
```

```
# only showing dimensions 1 and 2
#$`Dim 1`
#  trt      means M
#1   2  1.6394565 a
#2   3 -0.5558584 b
#3   1 -0.7550254 b

$`Dim 2`
  trt      means M
1   1  1.1357427 a
2   2  0.5321577 b
3   3 -0.8665043 c

#Since we want to project the IPM cluster means into the EPM space we
# need to do some manipulation of the data frame

#transpose cons to add cluster numbers
constrans<-t(cons)
head(constrans)
#            W1      W10      W12      W15       W16      W18      W19
#J10   5.8538507 5.645779 6.217736 6.898470 5.0134843 5.206546 6.538205 ..
#J100  6.8000000 5.529788 6.292733 7.052910 5.8000000 4.800000 5.900000 ..
#J101  1.7000000 2.936924 2.245496 2.136123 4.2462748 3.133009 4.100000 ..
#J102  0.6535121 3.199202 2.262261 2.154773 4.3000000 3.445800 4.696394 ..
#J103  4.1588675 3.787925 5.032193 8.200000 2.4409137 2.000000 5.839021 ..
#J104  4.9000000 2.649175 3.972374 4.700000 0.7350065 2.700000 2.978192 ..
#           W25        W26       W5        W9
#J10   5.442603  3.5000000 5.051085 5.2745977
#J100  5.007327  4.4443512 4.973365 4.7146176
#J101  2.800000  0.2268761 2.512882 4.9300475
#J102  5.245464 -0.7326598 0.300000 7.0000000
#J103  3.409422  0.2427217 2.434379 3.0612573
#J104  1.174173  2.8989304 2.220033 0.5113029

#now we create a new data set containing the clusters
clus= cbind(as.data.frame(constrans), clust=as.factor(cutree(clus.ward, k
= 3)))
clus$judge=row.names(clus)
head (clus)
#            W1      W10      W12      W15       W16      W18      W19
#J10   5.8538507 5.645779 6.217736 6.898470 5.0134843 5.206546 6.538205 ..
#J100  6.8000000 5.529788 6.292733 7.052910 5.8000000 4.800000 5.900000 ..
#J101  1.7000000 2.936924 2.245496 2.136123 4.2462748 3.133009 4.100000 ..
#J102  0.6535121 3.199202 2.262261 2.154773 4.3000000 3.445800 4.696394 ..
#J103  4.1588675 3.787925 5.032193 8.200000 2.4409137 2.000000 5.839021 ..
#J104  4.9000000 2.649175 3.972374 4.700000 0.7350065 2.700000 2.978192 ..
#           W25        W26       W5        W9 clust judge
#J10   5.442603  3.5000000 5.051085 5.2745977     1   J10
#J100  5.007327  4.4443512 4.973365 4.7146176     1  J100
#J101  2.800000  0.2268761 2.512882 4.9300475     2  J101
#J102  5.245464 -0.7326598 0.300000 7.0000000     2  J102
#J103  3.409422  0.2427217 2.434379 3.0612573     1  J103
#J104  1.174173  2.8989304 2.220033 0.5113029     1  J104

consMelt <- melt(clus, id.vars = 13:14)
colnames(consMelt)[3]="wine"
head(consMelt)
```

```
#  clust judge wine     value
#1     1   J10   W1 5.8538507
#2     1  J100   W1 6.8000000
#3     2  J101   W1 1.7000000
#4     2  J102   W1 0.6535121
#5     1  J103   W1 4.1588675
#6     1  J104   W1 4.9000000

# now we get the mean values for each cluster
clusters<-acast(consMelt, wine~clust, value.var= "value", fun.aggregate=me
an)
clusters
#           1        2        3
#W1   5.805209 2.222700 5.334765
#W10 4.732629 4.609450 4.677857
#W12 5.903607 3.533891 4.485406
#W15 7.247325 3.418364 4.020161
#W16 2.997200 6.333328 4.436407
#W18 3.723858 4.708253 4.765954
#W19 5.783650 5.729655 3.289039
#W20 2.957060 3.036443 6.328082
#W25 3.630929 6.426200 4.002938
#W26 3.363646 1.272416 7.150254
#W5   3.781959 3.997054 5.163112
#W9   3.194064 7.169784 3.942754

#we create a data set containing the DA mean data and the cluster means
epm<-cbind(DAmean, clusters)
# we perform the EPM, see Figure 3a and 3b and Figures 4a and 4b.
mdpref<-PCA(epm, scale = FALSE, quanti.sup=(ncol(DAmean)+1):ncol(epm))
mdpref<-PCA(epm, scale = FALSE, axes= c(1,3),quanti.sup=(ncol(DAmean)+1):n
col(epm))

# if we had wanted to use all the consumers instead of the cluster means
# we would have done the following (results not shown)
epm1<-cbind(DAmean, cons)
epm1
mdpref<-PCA(epm1, scale = FALSE, quanti.sup=(ncol(DAmean)+1):ncol(epm1))
mdpref<-PCA(epm1, scale = FALSE, axes= c(1,3),quanti.sup=(ncol(DAmean)+1):
ncol(epm1))
```

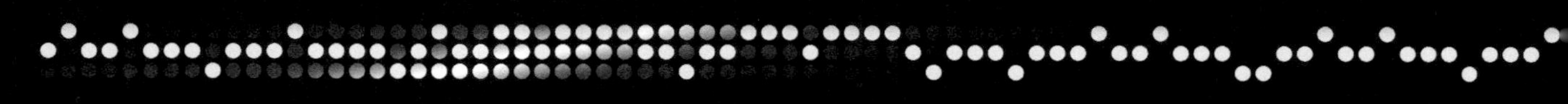

In this chapter

Objective

We were interested in determining how changes in chemical properties during storage at different temperatures change the sensory attributes of a wine. Additionally, we wanted to determine whether the closure type changed these effects.

Case Study 10

The effects of closure type and storage temperature on the sensory and chemical properties of a Cabernet Sauvignon wine[a]

10.1 Materials and methods

A Cabernet Sauvignon (vintage 2009) from the Californian Central Coast American Viticultural Area (AVA) was used for this study. The wine was bottled in 750 mL green glass bottles and closed with one of three closures [natural corks (naco; naco = 24 × 49 mm^2 AC-1 grade natural cork, ACI Cork, Fairfield, California, USA), synthetic cork (syco; syco = 22.5 × 43 mm^2 classic + synthetic cork, Nomacorc LLC, Zebulon, North Carolina, USA) or screw caps (screw; screw = 30 × 60 mm^2 aluminum Stelvin cap, Federfin Tech S.R.L., Tromello, Italy, with 28.6 × 2 mm^2 tin-PVDC liner, Oenoseal, Chazay D'Azergues, France)], then subsets of the wines were stored for 6 months at one of three storage temperatures (10, 20, 40°C), resulting in nine treatments. After the storage period the pH, titratable acidity (TA), volatile acidity (VA), free sulfur dioxide (fSO_2), and total sulfur dioxide (tSO_2) contents were measured as described in Hopfer et al. (2012).

Descriptive analysis (DA) was performed by a trained panel of 10 individuals. During five 1-h training sessions the panelists created, agreed on, and were trained on 38 aromas, flavors, tastes, and mouthfeel attributes. Table CS10.1 lists the reference standards for the attributes significant for the subset of the data used in this case study. Six wine samples were served in each session and each wine was evaluated in triplicate by each panelist. Panelists received 25 mL wine in clear pear-shaped ISO glasses (Anonymous, 1977) labeled with three digit random numbers. All samples were expectorated and filtered water and crackers were available as palate cleansers. Data were collected by FIZZ (Biosystèmes,

---

[a] *Based on data published by Hopfer et al. (2013).*

Sensory and Instrumental Evaluation of Alcoholic Beverages. http://dx.doi.org/10.1016/B978-0-12-802727-1.00016-8

Table CS10.1 Reference standards used in the descriptive analysis (DA)

Attribute	Reference standard recipe	
Red fruit	1 T Black cherry juice concentrate (RW Knudsen, Chico, CA) + 1 frozen strawberry (Dole, West Village, CA) + 5 g cut Roma tomatoes (0.5 × 0.5 cm^2 pieces) + 10 halved fresh raspberries (Driscoll's, Watsonville, CA) + 6 halved dried cranberries (Mariani Premium, Vacaville, CA) + 2 dried Bing cherries cut in six pieces (Trader Joe's, Monrovia, CA) in 20 mL base wine[a]	
Grapefruit	2 × 1 cm^2 Fresh grapefruit peel	
Canned veg	5 mL Brine cut asparagus with tips (Sunny Select, Walnut Creek, CA) + 5 mL brine Libby's Whole Kernel Sweet Corn (Seneca Foods Co., Marion, NY) + 5 mL brine green giant cut green beans (Minneapolis, MN) in 20 mL base wine	
Earthy	2 g All-purpose potting soil (Black Gold, Bellevue, WA) + 10 drops water	
Wood[b]	Woody	0.1 g EvOak American oak powder (Oak Solutions, Napa, CA) + 0.1 g EvOak French oak medium toast large chips (Oak Solutions, Napa, CA) in 25 mL base wine
	Cedar	One cedar ball (Cedar Fresh, Household Essentials, Hazelwood, MO)
Molasses/soy sauce	5 mL Soy sauce (Kikkoman, San Francisco, CA) + 1.5 g molasses from blue Agave in 40 mL base wine	
Oxidized[b]	Honey	3.5 g Honey (Nugget, Woodland, CA) in 25 mL base wine
	Sherry	10 mL Domecq Light Sherry Manzanilla
	Port wine	10 mL Prager Petit Sirah Port Lodi
Floral	0.5 g Nivea crème (Beiersdorf, Wilton, CT) + 0.1 mL violet essential oil (Aroma Crafts, Uttaranchal, India) solution (2 drops in 10 mL water)	
Hot mouthfeel (hotMF)	8% (v/v) Ethanol (Vodka 80 proof) in water	

[a] *As base wine Franzia Cabernet Sauvignon (Ripon, CA) was used unless otherwise noted.*
[b] *For this reference, more than one standard was available, but panelists rated only the combined attribute.*
All attributes were anchored with the words "low" and "high," with the exception of viscous ("thin" and "thick").

Couternon, France) and there was an enforced 1-min break between each wine with a 3-min break between wines three and four.

The Harbertson-Adams assay (Heredia et al., 2006) was used to determine total iron-reactive phenolics (IRP), total tannins (totTan) as (+)-catechin equivalents, small and large polymeric pigments (SPP, LPP, respectively) and total anthocyanins (totA). Volatile compound analyses were performed using headspace-solid phase microextraction with gas chromatography–mass spectrometry (HS-SPME-GCMS) using the method of Hjelmeland et al. (2012) with the additional measurement of diethyl succinate. Internal standard normalized areas of selected ion monitoring ions were used for further analyses and to compare relative changes among stored wines. Table CS10.2 lists all detected compounds with their identification features.

Table CS10.2 List of detected volatile compounds with their identification features using the method by Hjelmeland et al. (2012)

Compound	CAS #	RI_{lit}[a]	RI_{calc}	SIM ions
Ethyl acetate	141-78-6	907	915	43, 61, 88
Ethyl-2-methyl propanoate	97-62-1	955	960	43, 71, 116
Ethyl butanoate	105-54-4	1028	1022	116, 88, 71
Ethyl 2-methylbutanoate	7452-79-1	1050	1038	57, 102, 130
Ethyl 3-methylbutanoate	108-64-5	1069	1055	85, 88, 130
Methyl-2-propanol	78-83-1	1099	1101	43, 74, 55
Isoamyl acetate	123-92-2	1132	1126	55, 87, 130
Methyl-3-propanol	123-51-3	1205	1216	57, 70, 88
Ethyl hexanoate	123-66-0	1220	1238	88, 99, 144
Hexyl acetate	142-92-7	1270	1278	43, 84, 144
Hexanol	111-27-3	1360	1366	56, 69, 102
Ethyl octanoate	106-32-1	1436	1443	88, 101, 172
1-Octen-3-ol	3391-86-4	1449	1464	57, 72, 128
Vitispirane 1		1515	1526	177, 192, 93
Vitispirane 2			1529	177, 192, 93
2-Undecanone (IS)	112-12-9	1598	1604	58, 71, 170
Ethyl decanoate	110-38-3	1636	1645	88, 101, 200
Diethyl succinate	123-25-1	1666	1689	56
2-Phenethyl acetate	103-45-7	1829	1814	91, 104, 121
2-Phenethyl alcohol	60-12-8	1925	1916	65, 103, 122
Acetic acid[b]	64-19-7	1450	1461	
Isovaleric acid[b]	503-74-2	1665	1703	
Octanoic acid[b]	124-07-2	2083	2086	
Decanoic acid[b]	334-48-5	2361	2253	

[a]*Literature values RI on a DB-WAX column were obtained from Acree and Arn (n.d.), El-Sayed (n.d.), and Humpf and Schreier (1991).*

[b]*Areas for acids were obtained from the total ion chromatogram (TIC).*

RI, Retention index; SIM, selected ion monitoring.

Two bottles of each treatment were equipped with two noninvasive oxygen sensor spots each; the sensors measured the headspace (HS) and the dissolved (DO) oxygen throughout the storage period (5 mm sensor spots PSt3, NomaSense, Nomacorc LLC, Zebulon, North Carolina, USA). Oxygen levels were checked 21 times during the 6-month storage period. The limits of detection were 0.31 hPa for HS values and 15 ppb (= 0.015 ppm) for DO oxygen according to the manufacturer's specifications. HS values were measured in hPa and %oxygen, and DO oxygen was measured in ppm. Total packaged oxygen (TO) in ppm was calculated from the HS and the DO values as the sum of the two, after the HS values were converted into ppm. For the bottle treatments, a constant HS volume of 4.71 mL was used. For the calculation of the area under curve (auc) values, bottle measurements were averaged over the duplicates, and reported as ppm per day (ppm/day). We had shown in Hopfer et al. (2012) that the auc values correlated well with treatment. Thus, samples with low oxygen consumption and/or ingress (e.g., stored at lower temperatures in screw-capped bottles) had a higher auc than samples stored at higher temperatures. The auc is therefore a single value expressing oxygen consumption.

10.2 Data analysis

Univariate analysis of variance (ANOVA) for each sensory attribute using a three-way fixed effect model with all two-way interactions was performed (data not shown). For chemical parameters we performed univariate two-way fixed effect ANOVAs (data not shown). All significance testing used an alpha level of 5%. The mean values of the significant sensory and chemical attributes were used in the multifactor analysis (MFA) and the three oxygen measurements (HS, DO, and TO) were used as supplemental variables.

10.3 Results

Fourteen of the 38 sensory attributes differed significantly across treatments (Table CS10.3). In most cases the wines stored at 40°C were different from the ones stored at lower temperatures. The canned veg., earthy, oxidized, and molasses/soy characteristics increased at higher temperatures whereas the red fruit, floral, and woody attributes decreased in intensity. All wet chemistry attributes differed significantly among the wines as did all Harbertson-Adams parameters (Table CS10.4). Similarly to the sensory attributes, the totTan, totA, fSO_2, tSO_2 concentrations, and short chain polymeric pigments decreased when the wines were stored at 40°C relative to the lower temperatures. The VA of the 40°C wines was higher than the values in wines stored at lower temperatures but only for the natural cork wines. The pH of the screw closure wines was lower than the pH for the cork (synthetic and natural) closure wines. Additionally, all volatile compounds except hexanol, 2-phenyl ethyl alcohol, acetic acid, isovaleric acid, and octanoic acid differed significantly across treatments (Table CS10.5). Ten of

Table CS10.3 Mean sensory attribute values for wines[a] with Fisher's protected least significant difference (lsd) value

	Aroma compounds[b]					
	Red fruit	**Canned veg**	**Earthy**	**Molasses/ soy sauce**	**Oxidized**	**Floral**
naco10	3.0 ab	1.1 c	1.2 cd	1.1 c	1.4 bcd	1.8 a
naco20	3.2 a	1.1 c	1.3 cd	1.1 c	1.5 bcd	1.7 ab
naco40	2.4 bc	2.7 b	2.7 a	2.6 b	1.9 ab	0.9 c
screw10	3.0 ab	1.5 c	1.8 bc	1.4 c	1.8 abc	1.5 abc
screw20	2.9 ab	1.3 c	1.9 bc	1.6 c	1.1 cd	1.5 abc
screw40	2.0 c	3.9 a	2.8 a	3.0 ab	2.5 a	0.9 c
syco10	3.2 ab	1.1 c	1.6 bcd	1.5 c	1.0 d	1.4 abc
syco20	2.6 abc	1.000	1.0 d	1.0 c	1.2 bcd	2.1 a
syco40	2.5 abc	2.8 b	2.1 ab	3.7 a	2.4 a	1.0 bc
lsd	0.83	0.92	0.72	0.78	0.73	0.79

	Flavor compounds							Mouthfeel
	Red fruit	**Grape fruit**	**Canned veg**	**Earthy**	**Woody**	**Molasses/ soy sauce**	**Oxidized**	**Hot**
naco10	3.3 ab	1.7 abc	0.9 bcd	0.9 bc	1.4 bcd	1.3 bc	1.2 c	3.3 c
naco20	2.3 d	1.6 bc	0.5 d	0.8 c	1.4 bcd	1.8 b	1.2 c	4.5 ab
naco40	2.2 d	1.5 bc	1.4 abc	1.7 a	2.0 abc	2.0 b	2.0 ab	3.7 bc
screw10	3.1 abc	1.8 abc	0.6 d	1.2 bc	1.7 bcd	1.5 bc	1.4 bc	4.2 abc
screw20	2.5 bcd	2.4 a	0.7 cd	1.0 bc	2.5 a	1.5 bc	1.3 c	4.1 abc
screw40	2.1 d	1.3 c	1.7 a	1.3 ab	2.1 ab	2.0 b	2.6 a	4.1 abc
syco10	3.7 a	2.1 ab	0.6 d	1.0 bc	2.1 ab	0.9 c	1.2 c	3.3 c
syco20	3.2 ab	1.8 abc	0.9 bcd	1.0 bc	1.3 cd	1.5 bc	0.9 c	4.8 a
syco40	2.4 cd	2.3 a	1.6 ab	1.3 ab	1.3 d	2.8 a	2.3 a	3.8 bc
lsd	0.78	0.73	0.72	0.49	0.77	0.75	0.71	0.88

[a]*naco, Natural cork; syco, synthetic cork; screw, screw cap; and 10, 20, 40 is equal to 10, 20, and 40°C, respectively.*
[b]*Columns sharing the same letter are not significantly different from each other ($p < 0.05$).*
Letters indicate significant differences at $p<0.05$.

Table CS10.4 Mean chemical attribute values for wines[a] with Fisher's protected lsd value

	totTa[b]	totA	LPP	SPP	IRP
naco10	399.667 b	435.333 a	1.140 c	1.813 cd	1120.3 abc
naco20	365.667 d	406.333 b	1.160 c	1.737 d	1120.3 abc
naco40	327.000 ef	147.667 de	1.720 a	1.900 bc	1088.7 d
screw10	418.667 a	430.333 a	0.980 d	1.810 cd	1136.7 ab
screw20	375.333 cd	385.000 c	1.003 cd	1.833 c	1143.0 a
screw40	314.667 f	154.000 d	1.423 b	2.123 a	1117.0 bc
syco10	387.667 bc	434.000 a	0.933 d	1.987 b	1100.3 cd
syco20	364.667 d	390.333 bc	0.963 d	1.967 b	1115.3 bc
syco40	336.000 e	131.667 e	1.590 a	2.203 a	1082.0 d
lsd	15.37	19.78	0.16	0.09	25.04

	EtOH	pH	TA	fSO_2	tSO_2	VA
naco10	13.07 b	3.71 cde	6.17 d	31.3 a	106.0 b	0.47 c
naco20	13.06 b	3.71 bcd	6.13 e	24.0 b	106.3 b	0.48 c
naco40	13.03 e	3.72 abc	6.04 f	5.3 c	74.0 cd	0.61 ab
screw10	13.08 a	3.70 f	6.24 b	28.3 ab	129.0 a	0.65 a
screw20	13.04 d	3.70 ef	6.22 c	29.0 ab	111.3 b	0.65 a
screw40	13.06 bc	3.70 def	6.13 e	7.7 c	82.0 c	0.64 ab
syco10	13.09 a	3.72 abc	6.28 a	30.0 a	109.0 b	0.65 ab
syco20	13.08 a	3.72 ab	6.25 b	23.7 b	104.0 b	0.63 ab
syco40	13.05 c	3.73 a	6.15 d	3.7 c	68.7 d	0.60 b
lsd	0.009	0.011	0.021	5.57	13.01	0.045

[a]*naco, Natural cork; syco, synthetic cork; screw, screw cap; and 10, 20, 40 is equal to 10, 20, and 40°C, respectively; totTan, total tannin in mg/L; totA, total anthocyanin in mg/L; LPP, long polymeric pigments; SPP, short polymeric pigments; IRP, iron reactive pigment in mg/L; EtOH, ethanol as % (v/v), TA, titratable acidity in g/L;* fSO_2 *and* tSO_2*, free sulfur dioxide and total sulfur dioxide, respectively in mg/L, and VA, volatile acidity in g/L.*
[b]*Columns sharing the same letter are not significantly different from each other* ($p < 0.05$).
Letters indicate significant differences at $p<0.05$.

the 18 significant volatile compounds increased in wines stored at 40°C relative to wines stored at 20 or 10°C, with vitispirane 1 and vitispirane 2 being not detectable in the wines stored at the lower temperatures but very evident in the wine stored at 40°C. Additionally, screw capped wines stored at 40°C had the highest vitispirane concentrations (this may be consistent with enhanced

Table CS10.5 Mean volatile compound values for wines[a,b] with Fisher's protected lsd value

	Ethyl acetate	Ethyl-2-methyl propanoate	Ethyl butanoate	Ethyl-2-methyl butanoate	Ethyl-3-methyl butanoate	Methyl-2-propanol
naco10	443 e	9.3 f	9.3 d	3.4 f	4.8 g	19.5 cde
naco20	459 e	10.8 e	9.1 d	4.0 e	5.6 e	18.3 e
naco40	593 c	17.7 c	10.0 bc	8.0 c	11.2 c	19.3 de
screw10	466 e	11.7 e	10.1 bc	3.7 e	5.3 ef	20.8 bc
screw20	513 d	13.3 d	10.15 b	4.5 d	6.3 d	21.3 ab
screw40	636 b	18.9 b	11.0 a	8.3 b	11.7 b	21.4 ab
syco10	449 e	11.1 e	9.5 cd	0.0 g	5.0 fg	20.1 bcd
syco20	512 d	13.3 d	10.1 bc	4.5 d	6.3 d	21.1 ab
syco40	687 a	20.7 a	11.5 a	9.3 a	13.0 a	22.5 a
	32.90	0.98	0.58	0.30	0.42	1.39
	Isoamyl acetate	**Methyl-3-propanol**	**Ethyl hexanoate**	**Hexyl acetate**	**Ethyl octanoate**	**1-Octen-3-ol**
naco10	50.5 b	371.9 bcd	180.4 c	7.5 b	471.2 c	0.0 f
naco20	42.5 d	346.9 d	183.6 c	6.4 d	505.2 b	4.8 e
naco40	26.6 f	365.8 cd	200.0 b	0.0 e	432.7 d	29.2 b
screw10	54.1 a	387.8 bc	200.0 b	8.4 a	565.3 a	0.0 f
screw20	47.1 c	400.0 ab	198.6 b	7.0 c	572.6 a	5.6 de
screw40	30.7 e	398.6 ab	221.8 a	0.0 e	572.4 a	26.2 c
syco10	50.9 b	381.2 bc	167.7 d	7.1 c	331.2 e	0.0 f
syco20	46.0 c	397.6 ab	177.9 cd	6.2 d	343.2 e	5.8 d
syco40	30.5 e	420.9 a	197.1 b	0.0 e	227.4 f	32.0 a
	2.82	29.85	11.10	0.36	23.65	0.87
	Vitispirane1	**Vitispirane2**	**Ethyl decanoate**	**Diethyl succinate**	**2-Phenylethyl acetate**	**Decanoic acid**
naco10	0.0 d	0.0 d	124.8 d	169.0 d	9.3 ab	0.0 f
naco20	0.0 d	0.0 d	129.2 c	173.8 cd	8.5 c	1.5 cd
naco40	20.3 b	11.0 b	74.0 f	234.8 b	5.6 e	1.5 cd
screw10	0.0 d	0.0 d	175.5 a	171.5 cd	9.7 a	1.6 bc
screw20	0.0 d	0.0 d	173.9 a	186.6 cd	9.2 b	2.0 a
screw40	21.6 a	11.6 a	154.7 b	234.7 b	6.7 d	1.8 ab
syco10	0.0 d	0.0 d	96.3 e	175.5 cd	9.2 ab	1.1 e
syco20	0.0 d	0.0 d	93.6 e	192.2 c	9.0 b	1.1 e
syco40	10.8 c	6.0 c	45.8 g	267.9 a	6.4 d	1.3 de
	0.39	0.22	4.02	21.50	0.42	0.25

[a] *naco, Natural cork; syco, synthetic cork; screw, screw cap; and 10, 20, 40 is equal to 10, 20, and 40°C, respectively.*
[b] *Columns sharing the same letter are not significantly different from each other (p < 0.05).*
Letters indicate significant differences at $p<0.05$.

Table CS10.6 The mean area under the curve (auc) of headspace (HS), dissolved (DO), and total (TPO) oxygen concentrations (ppm/day) in the bottles with different closures and different storage temperatures[a]

	HS	DO	TPO
naco10	23.5	38.78	62.29
naco20	22.27	15.35	37.63
naco40	15.80	15.91	31.71
screw10	11.90	40.30	52.20
screw20	19.69	30.84	50.53
screw40	9.04	26.23	35.27
syco10	26.04	37.02	63.08
syco20	26.19	14.55	40.75
syco40	22.22	16.36	38.58

[a]*naco, Natural cork; syco, synthetic cork; screw, screw cap; and 10, 20, 40 is equal to 10, 20, and 40°C, respectively.*

hydrolysis of glycoside precursors at the higher storage temperatures). Four volatiles compound, all esters (ethyl hexanoate, hexyl acetate, ethyl decanoate, and 2-phenylethyl acetate) decreased in wines stored at 40°C relative to wines stored at lower temperatures (this may be associated with enhanced ester hydrolysis at the higher storage temperature). The same was true for the ethyl octanoate concentration in cork closed wines, whether synthetic or not, but in the case of the screw-capped wines the concentration of ethyl octanoate did not change significantly with temperature. Table CS10.6 shows the auc values for the HS, DO, and TO in the bottles of each treatment. It is clear that wines with little or no oxygen ingress (screw) or wines stored at lower temperatures had higher auc values than wines with higher oxygen ingress (naco40). This means that less of the present oxygen was consumed during oxidative reactions.

Fig. CS10.1 shows the MFA space and it is clear in Fig. CS10.1A that the horizontal axis separates the wines stored at 40°C from those stored at 10 or 20°C. The 95% confidence ellipses in Fig. CS10.1B show that the synthetic cork capped wines stored at 10 and 20°C or at 40°C differed from the screw capped wines stored at 10 and 20°C or at 40°C, respectively. The natural cork closed wines did not differ from either the synthetic cork closed or the screw cap wines.

Fig. CS10.1C shows the loadings for all the variables used in the MFA as well as the locations of the oxygen supplemental variables. However, it is very difficult to interpret this figure, therefore we replotted the same information with only the top 15 contributing variables (Fig. CS10.2A) as well as the figure with the top 15 coordinates (Fig. CS10.2B). Variable contributions to a dimension (component)

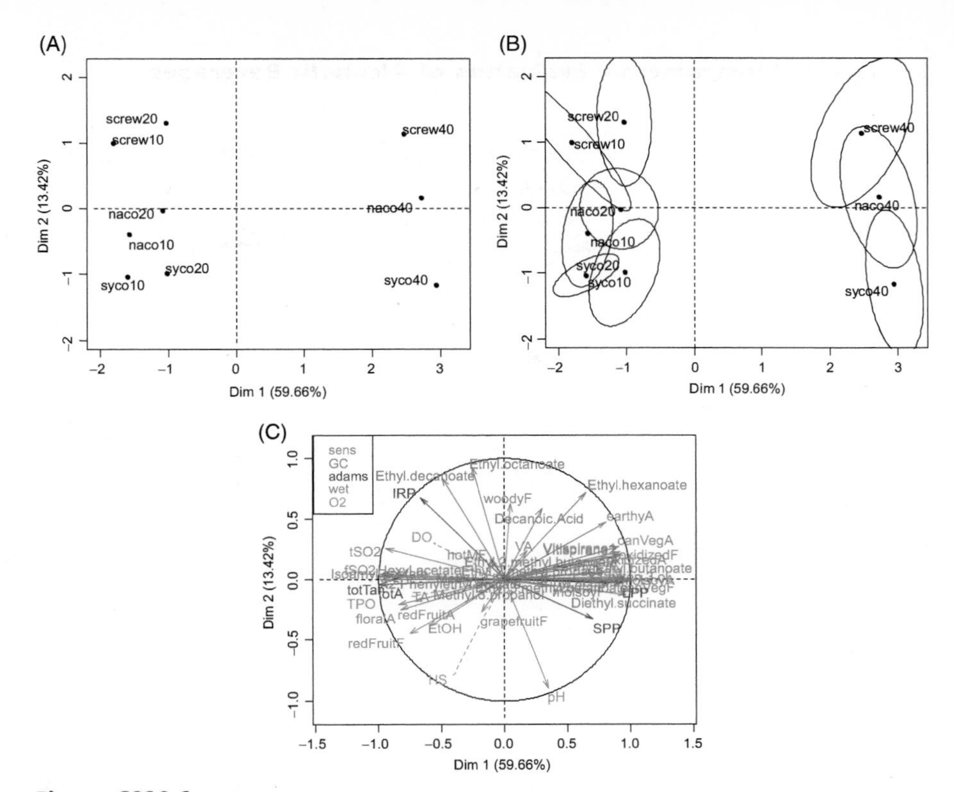

Figure CS10.1

(A) Multifactor analysis (MFA) plot of nine wine samples. Score plot for dimensions 1 and 2. Please see Table CS10.3 for codes. (B) MFA plot of nine wine samples. Score plot with 95% confidence ellipses for dimensions 1 and 2. Please see Table CS10.3 for codes. (C) MFA plot of nine wine samples. Loadings plot for dimensions 1 and 2. Please see Tables CS10.3–CS10.6 for codes. The supplemental variables are indicated by *dashed vector lines*.

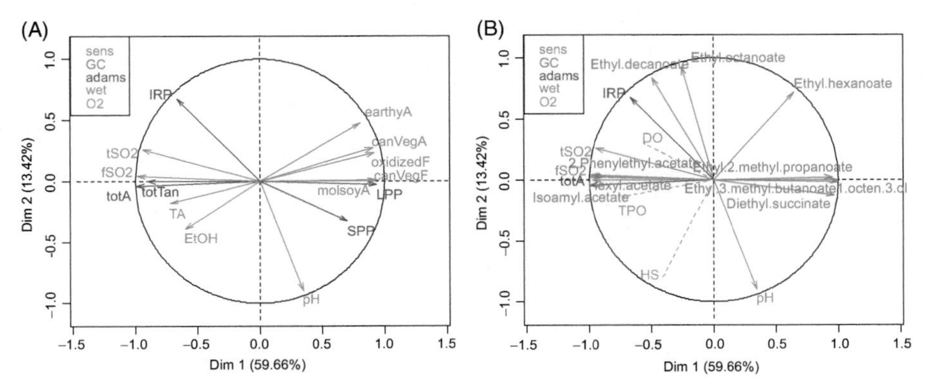

Figure CS10.2

(A) MFA plot of nine wine samples. Loadings plot of the top 15 contributing variables for dimensions 1 and 2. Please see Tables CS10.3–CS10.6 for codes. (B) MFA plot of nine wine samples. Loadings plot of the 15 variables with the highest coordinates for dimensions 1 and 2. Please see Tables CS10.3–CS10.6 for codes. The supplemental variables are indicated by *dashed vector lines*.

Table CS10.7 The list of variables significantly linked to dimensions 1 and 2 based on the MFA

	Correlation	*P* value
Dimension 1		
1-Octen-3-ol	0.997	4.03e-09
Ethyl-3-methyl butanoate	0.989	4.10e-07
Diethyl succinate	0.966	2.38e-05
Ethyl-2-methyl propanoate	0.925	7.39e-05
Ethyl acetate	0.947	1.05e-04
LPP	0.937	1.98e-04
Molasses/soy sauce aroma	0.936	2.02e-04
Vitispirane 2	0.926	3.33e-04
Vitispirane 1	0.921	4.20e-04
Canned veg flavor	0.921	4.22e-04
Ethyl-2-methyl butanoate	0.909	6.95e-04
Canned veg aroma	0.906	7.74e-04
Molasses/soy sauce flavor	0.827	5.94e-03
Oxidized aroma	0.813	7.76e-03
Earthy aroma	0.804	9.04e-03
Earthy flavor	0.790	1.12e-02
Ethyl butanoate	0.736	2.38e-02
SPP	0.699	3.60e-02
Iron reactive phenols	−0.668	4.93e02
TA	−0.721	2.81e-02
TPO	−0.739	2.30e-02
Red fruit flavor	−0.753	1.93e-02
Floral aroma	−0.829	5.70e-03
Red fruit aroma	−0.842	4.41e-03
totTan	−0.903	8.59e-04
tSO_2	−0.945	1.25e-05

Table CS10.7 The list of variables significantly linked to dimensions 1 and 2 based on the MFA (*cont.*)

	Correlation	*P* value
Isoamyl acetate	−0.970	1.50e-05
Phenethyl acetate	−0.975	7.54e-05
fSO_2	−0.985	1.28e-06
Hexyl acetate	−0.994	4.28e-06
Total anthocyanins	−0.998	5.01e-08
Dimension 2		
Ethyl octanoate	0.924	3.64e-04
Ethyl decanoate	0.836	4.96e-03
Ethyl hexanoate	0.715	3.05e-02
HS	−0.798	9.90e-03
pH	−0.898	1.00e-03

have a value between 0 and 1 and all variables associated with a component sum to 1. Thus, variables with high contribution on a dimension are more important to that dimension than one with lower contributions (Abdi et al., 2013). In the second case, the highest squared coordinates are chosen (Lê and Worch, 2015). Lastly, we also asked the MFA program to list the variables that were significantly linked to each dimension (Table CS10.7). Based on all the aforementioned we can say that TPO is associated with wines that had been stored at lower temperatures and thus less oxygen in the bottles was consumed. Wine stored at higher temperatures had lower sulfur dioxide values, lower total tannins and anthocyanins content as well as less phenethyl acetate. While wines stored at 40°C had more SPP and LPP, respectively, and showed canned veg aromas and flavors, oxidized aromas and flavors, and earthier aroma. Screw cap closed bottles had higher concentrations of ethyl decanoate, ethyl octanoate, and ethyl hexanoate than bottles closed with synthetic corks. The bottles with synthetic corks had higher pH and more HS oxygen.

10.4 Conclusions

It is clear that both storage temperature and closure type affect red wine sensory and chemical attributes, but storage at elevated temperatures, such as 40°C for 6 months affects in general wine composition and sensory properties more than the closure type.

References

Abdi, H., Williams, L.J., Valentin, D., 2013. Multiple factor analysis: principal component analysis for multitable and multiblock data sets. WIREs Comp. Stat. 5, 149–179.

Acree, T., Arn, H, n.d. Flavornet and human odor space gas chromatography—olfactometry (GCO) of natural products. Available from: http://www.flavornet.org

Anonymous, 1977. International Organization for Standardization ISO 3591: Sensory Analysis—Apparatus—Wine Tasting Glass.

El-Sayed, A.M., n.d. The Pherobase: database of pheromones and semiochemicals. Available from: http://www.pherobase.com

Heredia, T.M., Adams, D.O., Fields, K.C., Held, P.G., Harbertson, J.F., 2006. Evaluation of a comprehensive red wine phenolics assay using a microplate reader. Am. J. Enol. Vitic. 57, 497–502.

Hjelmeland, A.K., King, E.S., Ebeler, S.E., Heymann, H., 2012. Characterizing the chemical and sensory profiles of US Cabernet Sauvignon wines and blends. Am. J. Enol. Vitic. 64, 169–179.

Hopfer, H., Buffon, P.A., Ebeler, S.E., Heymann, H., 2013. The combined effects of storage temperature and packaging on the sensory, chemical and physical properties of a Cabernet Sauvignon wine. J. Agric. Food Chem. 61, 3320–3334.

Hopfer, H., Ebeler, S.E., Heymann, H., 2012. The combined effects of storage temperature and packaging type on the sensory and chemical properties of Chardonnay. J. Agric. Food Chem. 60, 10743–10754.

Humpf, H.U., Schreier, P., 1991. Bound aroma compounds from the fruit and the leaves of blackberry (*Rubus laciniata* L.). J. Agric. Food Chem. 39, 1830–1832.

Lê, S., Worch, T., 2015. Analyzing Sensory Data With R. CRC Press, Boca Raton, FL.

```
# all code comes without any warranty
# MFA for wine sensory and chemical parameters
# we imported a file named tempsign.csv
# 9 wines with 14 sensory attributes, 18 chemical attributes, 5
# Harbertson-Adams attributes, 6 wet chemistry and 3 oxygen attributes.
# The O2 attributes will be used as supplementary information.
library(SensoMineR)

# Import the data
allsign <- read.csv(file = "tempsign.csv", header = TRUE, sep = ",", dec =
".")
row.names(allsign)=allsign$wine
all<-allsign[,-1]

# Now we had to create the subsets for the MFA regression. One for sensory
# data (sens), and one each for the volatile, Harbertson-Adams, basic
# wine chemistry and O2 data. We also chose to create the 95% confidence
# ellipses around the mean data points in the score plot.

mfa.all <- MFA(all, group=c(14,18,5,6,3),  ncp=5,
name.group=c("sens","GC", "adams", "wet", "O2"), num.group.sup=5,
graph=FALSE)
prod= NULL
for(i in 1:nrow(all)){prod=c(prod, rep(row.names(all)[i], 4))}
ell.all=simule(cbind(as.factor(prod), mfa.all$ind$coord.partiel), 500)
ell12 = coord.ellipse(ell.all$simul, centre=NULL, axes=c(1,2),
level.conf=0.95)

summary(mfa.all)

#Eigenvalues
#               Dim.1   Dim.2   Dim.3   Dim.4   Dim.5   Dim.6   Dim.7
#Variance       3.735   0.840   0.755   0.302   0.237   0.199   0.111
#% of var.     59.656  13.422  12.061   4.824   3.786   3.177   1.769
#Cumulative %  59.656  73.078  85.140  89.963  93.749  96.926  98.696
#               Dim.8
#Variance       0.082
#% of var.      1.304
#Cumulative % 100.000

#Groups
#         Dim.1    ctr   cos2    Dim.2    ctr   cos2    Dim.3    ctr   cos2
#sens     0.915 24.508  0.772 |  0.136 16.238  0.017 |  0.062  8.170  0.004
#GC       0.972 26.021  0.866 |  0.211 25.150  0.041 |  0.180 23.780  0.030
#adams    0.955 25.574  0.877 |  0.149 17.765  0.021 |  0.114 15.090  0.012
#wet      0.893 23.897  0.627 |  0.343 40.848  0.093 |  0.400 52.960  0.126

#Supplementary group
#         Dim.1  cos2   Dim.2  cos2   Dim.3  cos2
#O2       0.544 0.220 | 0.396 0.116 | 0.057 0.002 |

#Individuals
#         Dim.1    ctr   cos2    Dim.2    ctr   cos2    Dim.3    ctr   cos2
#naco10  -1.570  7.332  0.526 | -0.392  2.028  0.033 | -1.067 16.740  0.243
#naco20  -1.088  3.520  0.285 | -0.031  0.012  0.000 | -1.514 33.721  0.551
#naco40   2.715 21.925  0.777 |  0.159  0.333  0.003 | -1.052 16.272  0.117
#screw10-1.801  9.650  0.606 |  0.993 13.038  0.184 |  0.663  6.460  0.082
#screw20-1.031  3.164  0.251 |  1.308 22.602  0.404 |  0.468  3.220  0.052
```

```
#screw40 2.463 18.042  0.709 |  1.142 17.237  0.152 |  0.491  3.548  0.028
#syco10 -1.601  7.623  0.481 | -1.030 14.019  0.199 |  0.767  8.659  0.110
#syco20 -1.022  3.109  0.301 | -0.987 12.875  0.280 |  0.612  5.511  0.108
#syco40  2.936 25.636  0.778 | -1.162 17.855  0.122 |  0.632  5.868  0.036

#Continuous variables (the 10 first)
#         Dim.1    ctr   cos2    Dim.2    ctr   cos2     Dim.3    ctr   cos2
#rdFrtA -0.842  2.169  0.709 | -0.206  0.576  0.042 | -0.188  0.537  0.035
#cnVegA  0.906  2.510  0.820 |  0.276  1.034  0.076 |  0.117  0.209  0.014
#earthy  0.804  1.977  0.646 |  0.474  3.050  0.224 |  0.098  0.146  0.010
#mlsoyA  0.936  2.682  0.877 |  0.008  0.001  0.000 |  0.221  0.741  0.049
#oxA     0.813  2.021  0.660 |  0.190  0.491  0.036 |  0.030  0.013  0.001
#flralA -0.829  2.104  0.688 | -0.245  0.817  0.060 | -0.118  0.212  0.014
#rdFrtF -0.753  1.733  0.566 | -0.450  2.754  0.203 |  0.308  1.436  0.095
#gpfrtF -0.184  0.104  0.034 | -0.271  0.998  0.073 |  0.458  3.175  0.210
#cnVegF  0.921  2.596  0.848 |  0.004  0.000  0.000 |  0.132  0.264  0.017
#earthF  0.790  1.912  0.625 |  0.066  0.059  0.004 |  0.095  0.138  0.009

#Supplementary continuous variables
#            Dim.1   cos2    Dim.2   cos2    Dim.3   cos2
#HS..ppm.d.  -0.408  0.166 | -0.798  0.637 | -0.085  0.007 |
#DO..ppm.d.  -0.561  0.314 |  0.302  0.091 |  0.253  0.064 |
#TPO..ppm.d.-0.739   0.546 | -0.136  0.019 |  0.192  0.037 |

#Then we asked for the RV coefficients of the groups
mfa.all$group$RV

#            sens        GC      adams       wet        O2       MFA
#sens   1.0000000 0.8901349 0.7820918 0.6860097 0.5206536 0.9016132
#GC     0.8901349 1.0000000 0.9102417 0.8006741 0.5407972 0.9678387
#adams  0.7820918 0.9102417 1.0000000 0.8330637 0.4450621 0.9479196
#wet    0.6860097 0.8006741 0.8330637 1.0000000 0.5550167 0.8978744
#O2     0.5206536 0.5407972 0.4450621 0.5550167 1.0000000 0.5565726
#MFA    0.9016132 0.9678387 0.9479196 0.8978744 0.5565726 1.0000000


#Then we plotted the loadings plot (choix="var"), score plot (choix="ind")
#and the score plot with the confidence ellipses for dimensions 1 and 2
(Figure 1a, 1b and 1c)
plot(mfa.all, axes=c(1,2), choix="ind")
plot(mfa.all, axes=c(1,2), choix="ind", ellipse=ell12)
plot(mfa.all, axes=c(1,2), choix="var")

# to simplify the explanation of figure1c we asked for a list of the
#variables significantly linked to each component or axis. This list is
#sorted by the sign of the correlation and the significance of the p-value
impvar<-dimdesc(mfa.all)
impvar$Dim.1
impvar$Dim.2

#Dim1
#                              correlation      p.value
#X1.octen.3.ol                   0.9971553 4.031342e-09
#Ethyl.3.methyl.butanoate        0.9893165 4.106911e-07
#Diethyl.succinate               0.9656870 2.382675e-05
#Ethyl.2.methyl.propanoate       0.9524039 7.392098e-05
# and so forth (see Table 7
```

```
#we could also plot the partial axes plot for dimensions 1 and 2 (Figure
not #shown)
plot.MFA(mfa.all, axes=c(1,2), choix="ind", partial="all",
habillage="group")
#or we could plot the group representation in dimensions 1 and 2 (Figure
#not shown)
plot.MFA(mfa.all, axes=c(1,2), choix="group", habillage="none")

#lastly, to facilitate interpretation we plotted the loadings plot with
#the top 15 contributing variables (Figure 2a) and with the top 15
#coordinates (Figure 2b)

plot(mfa.all, axes=c(1,2), choix="var", select="contrib 15")
plot(mfa.all, axes=c(1,2), choix="var", select="coord 15")
```

Index

D